Nuclear Chemistry

Nuclear Chemistry

Johnny Hofstader

Larsen & Keller
www.larsen-keller.com

Nuclear Chemistry
Johnny Hofstader
ISBN: 978-1-64172-631-3 (Hardback)

Larsen & Keller
Published by Larsen and Keller Education,
5 Penn Plaza,
19th Floor,
New York, NY 10001, USA

Cataloging-in-Publication Data

Nuclear chemistry / Johnny Hofstader.
p. cm.
Includes bibliographical references and index.
ISBN 978-1-64172-631-3
1. Nuclear chemistry. 2. Chemistry, Physical and theoretical. I. Hofstader, Johnny.
QD601.3 .N83 2022
541.38--dc23

For more information regarding Larsen and Keller Education and its products, please visit the publisher's website www.larsen-keller.com

Table of Contents

Preface

The branch of chemistry which deals with nuclear processes, radioactivity and transformations in the nuclei of atoms is called nuclear chemistry. Some of the transformations studied within it are nuclear transmutation and nuclear properties. It is also involved in the study of radioactive elements such as the actinides, radium and radon as well as the equipment that is designed to perform nuclear processes. The study of the chemical effects of the absorption of radiation in living animals, plants and other materials also falls under this field. The main areas that are covered under nuclear chemistry are radiation chemistry, nuclear power and nuclear reactions. This textbook provides comprehensive insights into the field of nuclear chemistry. Also included herein is a detailed explanation of the various concepts and applications of this field. This book aims to serve as a resource guide for students and experts alike and contribute to the growth of the discipline.

To facilitate a deeper understanding of the contents of this book a short introduction of every chapter is written below:

Chapter 1- The branch of chemistry which focuses on the study of radioactivity, nuclear processes transformations within the nuclei of atoms is called nuclear chemistry. Some of the other aspects which are dealt with in this field are nuclear structure, size, shape and density. This chapter has been carefully written to provide an easy understanding of these facets of nuclear chemistry.

Chapter 2- The branch of nuclear chemistry which deals with the study of chemical effects of radiation on matter is known as radiation chemistry. The techniques used in radiation chemistry include exposure to a steady radiation source and exposure to a pulsed radiation source. The topics elaborated in this chapter will help in gaining a better perspective about radiation chemistry and these techniques used within it.

Chapter 3- The domain of nuclear chemistry which studies the chemical reactions and properties of non-radioactive isotopes using the radioactive isotopes of elements is called radiochemistry. Radioactive decay refers to the process through which an unstable atomic nucleus loses energy by emitting radiation. This chapter discusses in detail the concepts and theories related to radiochemistry and radioactive decay such as radioactive isotope and resonance fluorescence.

Chapter 4- The process in which two nuclear particles interact and produce two or more nuclear particles is called a nuclear reaction. Some of the areas of study related to nuclear reactions are photonuclear reaction, energetics of nuclear reactions and compound nucleus reactions. These aspects of nuclear reactions have been thoroughly discussed in this chapter.

Chapter 5- Nuclear chemistry plays an integral role in the synthesis of thorium and uranium for use in nuclear reactors. It is also involved in the fabrication of nuclear fuel as well as in the treatment of radioactive waste. The chapter closely examines these key applications of nuclear chemistry in the field of nuclear power to provide an extensive understanding of the subject.

Chapter 6- The branch of medicine which seeks to apply radioactive substances for diagnosing and treating diseases is called nuclear medicine. Some of its applications include nuclear medicine imaging and radiopharmaceuticals. The topics elaborated in this chapter will help in gaining a better perspective about these aspects and applications of nuclear medicine.

I would like to share the credit of this book with my editorial team who worked tirelessly on this book. I owe the completion of this book to the never-ending support of my family, who supported me throughout the project.

Johnny Hofstader

1

An Introduction to Nuclear Chemistry

The branch of chemistry which focuses on the study of radioactivity, nuclear processes transformations within the nuclei of atoms is called nuclear chemistry. Some of the other aspects which are dealt with in this field are nuclear structure, size, shape and density. This chapter has been carefully written to provide an easy understanding of these facets of nuclear chemistry.

Nuclear chemistry is the study of the chemical and physical properties of elements as influenced by changes in the structure of the atomic nucleus. Modern nuclear chemistry, sometimes referred to as radiochemistry, has become very interdisciplinary in its applications, ranging from the study of the formation of the elements in the universe to the design of radioactive drugs for diagnostic medicine. In fact, the chemical techniques pioneered by nuclear chemists have become so important that biologists, geologists, and physicists use nuclear chemistry as ordinary tools of their disciplines. While the common perception is that nuclear chemistry involves only the study of radioactive nuclei, advances in modern mass spectrometry instrumentation has made chemical studies using stable, nonradioactive isotopes increasingly important.

There are essentially three sources of radioactive elements. Primordial nuclides are radioactive elements whose half-lives are comparable to the age of our solar system and were present at the formation of Earth. These nuclides are generally referred to as naturally occurring radioactivity and are derived from the radioactive decay of thorium and uranium. Cosmogenic nuclides are atoms that are constantly being synthesized from the bombardment of planetary surfaces by cosmic particles (primarily protons ejected from the Sun), and are also considered natural in their origin. The third source of radioactive nuclides is termed anthropogenic and results from human activity in the production of nuclear power, nuclear weapons, or through the use of particle accelerators.

Lasers focus on a small pellet of fuel in attempt to create a nuclear fusion reaction (the combination of two nuclei to produce another nucleus) for the purpose of producing energy.

Marie Curie was the founder of the field of nuclear chemistry. She was fascinated by Antoine-Henri Becquerel's discovery that uranium minerals can emit rays that are able to expose photographic film, even if the mineral is wrapped in black paper. Using an electrometer invented by her husband Pierre and his brother Jacques that measured the electrical conductivity of air (a precursor to the Geiger counter), she was able to show that thorium also produced these rays—a process that she called radioactivity. Through tedious chemical separation procedures involving precipitation of different chemical fractions, Marie was able to show that a separated fraction that had the chemical properties of bismuth and another fraction that had the chemical properties of barium were much more radioactive per unit mass than the original uranium ore. She had separated and discovered the elements polonium and radium, respectively. Further purification of radium from barium produced approximately 100 milligrams of radium from an initial sample of nearly 2,000 kilograms of uranium ore.

In 1911 Ernest Rutherford asked a student, George de Hevesy, to separate a lead impurity from a decay product of uranium, radium-D. De Hevesy did not succeed in this task (we now know that radium-D is the radioactive isotope ^{210}Pb), but this failure gave rise to the idea of using radioactive isotopes as tracers of chemical processes. With Friedrich Paneth in Vienna in 1913, de Hevesy used ^{210}Pb to measure the solubility of lead salts—the first application of an isotopic tracer technique. De Hevesy went on to pioneer the application of isotopic tracers to study biological processes and is generally considered to be the founder of a very important area in which nuclear chemists work today, the field of nuclear medicine. De Hevesy also is credited with discovering the technique of neutron activation analysis, in which samples are bombarded by neutrons in a nuclear reactor or from a neutron generator, and the resulting radioactive isotopes are measured, allowing the analysis of the elemental composition of the sample.

In Germany in 1938, Otto Hahn and Fritz Strassmann, skeptical of claims by Enrico Fermi and Irène Joliot-Curie that bombardment of uranium by neutrons produced new so-called transuranic elements (elements beyond uranium), repeated these experiments and chemically isolated a radioactive isotope of barium. Unable to interpret these findings, Hahn asked Lise Meitner, a physicist and former colleague, to propose an explanation for his observations. Meitner and her nephew, Otto Frisch, showed that it was possible for the uranium nucleus to be split into two smaller nuclei by the neutrons, a process that they termed "fission." The discovery of nuclear fission eventually led to the development of nuclear weapons and, after World War II, the advent of nuclear power to generate electricity. Nuclear chemists were involved in the chemical purification of plutonium obtained from uranium targets that had been irradiated in reactors. They also developed chemical separation techniques to isolate radioactive isotopes for industrial and medical uses from the fission products wastes associated with plutonium production for weapons. Today, many of these same chemical separation techniques are being used by nuclear chemists to clean up radioactive wastes resulting from the fifty-year production of nuclear weapons and to treat wastes derived from the production of nuclear power.

In 1940, at the University of California in Berkeley, Edwin McMillan and Philip Abelson produced the first manmade element, neptunium (Np), by the bombardment of uranium with low energy neutrons from a nuclear accelerator. Shortly thereafter, Glenn Seaborg, Joseph Kennedy, Arthur Wahl, and McMillan made the element plutonium by bombarding uranium targets with deuterons, particles derived from the heavy isotope of hydrogen, deuterium (^2H). Both McMillan and Seaborg recognized that the chemical properties of neptunium and plutonium did not resemble those of rhenium and osmium, as many had predicted, but more closely resembled the chemistry of uranium, a fact that led Seaborg in 1944 to propose that the transuranic elements were part of a new group of elements called the actinide series that should be placed below the lanthanide series on the periodic chart. Seaborg and coworkers went on to discover many more new elements and radioactive isotopes and to study their chemical and physical properties. At the present, nuclear chemists are involved in trying to discover new elements beyond the 112 that are presently confirmed and to study the chemical properties of these new elements, even though they may exist for only a few thousandths of a second.

Nobel laureate Glenn T. Seaborg was among those who discovered many radioactive elements and isotopes.

As early as 1907 Bertram Boltwood had used the discovery of radioactive decay laws by Ernest Rutherford and Frederick Soddy to ascribe an age of over two billion years to a uranium mineral. In 1947 Willard Libby at the University of Chicago used the decay of ^{14}C to measure the age of dead organic matter. The cosmogenic radionuclide, ^{14}C, becomes part of all living matter through photosynthesis and the consumption of plant matter. Once the living organism dies, the ^{14}C decays at a known rate, enabling a date for the carbon-containing relic to be calculated. Today, scientists ranging from astrophysicists to marine biologists use the principles of radiometric dating to study problems as diverse as determining the age of the universe to defining food chains in the oceans. In addition, newly developed analytical techniques such as accelerator mass spectrometry (AMS) have allowed nuclear chemists to extend the principles of radiometric dating to nonradioactive isotopes in order to study modern and ancient processes that are affected by isotopic fractionation. This isotopic fractionation results from temperature differences in the environment in which the material was formed (at a given temperature, the lighter isotope will be very slightly more reactive than the heavier isotope), or from different chemical reaction sequences.

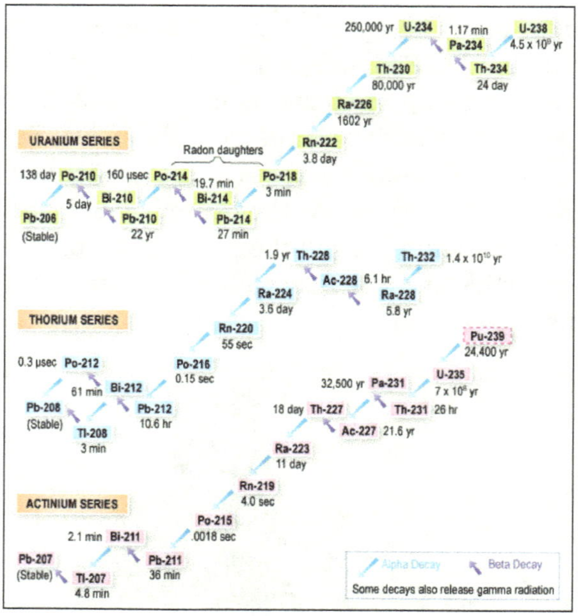

Naturally occurring radionuclides.

The newest area in which nuclear chemists play an important role is the field of nuclear medicine. Nuclear medicine is a rapidly expanding branch of health care that uses short-lived radioactive isotopes to diagnose illnesses and to treat specific diseases. Nuclear chemists synthesize drugs from radionuclides produced in nuclear reactors or accelerators that are injected into the patient and will then seek out specific organs or cancerous tumors. Diagnosis involves use of the radiopharmaceutical to generate an image of the tumor or organ to identify problems that may be missed by x rays or physical examinations. Treatment involves using radioactive compounds at carefully controlled doses to destroy tumors. These nuclear medicine techniques hold much promise for the future because they use biological chemistry to specify target cells much more precisely than traditional radiation therapy, which uses radiation from external sources to kill tumor cells, killing nontarget cells as well. Additionally, the use of nuclear pharmaceuticals containing the short-lived isotope ^{11}C has allowed nuclear chemists and physicians to probe brain activity to better understand the biochemical basis of illnesses ranging from Parkinson's disease to drug abuse.

Nuclear Structure and Stability

Nuclear chemistry is the study of reactions that involve changes in nuclear structure. The nucleus of an atom is composed of protons and, with the exception of $^1_1$H, neutrons. The number of protons in the nucleus is called the atomic number (Z) of the element, and the sum of the number of protons and the number of neutrons is the mass number (A). Atoms with the same atomic number but different mass numbers are isotopes of the same element. When referring to a single type of nucleus, we often use the term nuclide and identify it by the notation $^A_Z X$, where X is the symbol for the element, A is the mass number, and Z is the atomic number (for example, $^{14}_{6}\mathrm{C}$). Often a nuclide is referenced by the name of the element followed by a hyphen and the mass number. For example, $^A_Z X$ is called "carbon-14."

Protons and neutrons, collectively called nucleons, are packed together tightly in a nucleus. With a radius of about 10^{-15} meters, a nucleus is quite small compared to the radius of the entire atom, which is about 10^{-10} meters. Nuclei are extremely dense compared to bulk matter, averaging 1.8×10^{14} grams per cubic centimeter. For example, water has a density of 1 gram per cubic centimeter, and iridium, one of the densest elements known, has a density of 22.6 g/cm^3. If the earth's density were equal to the average nuclear density, the earth's radius would be only about 200 meters (earth's actual radius is approximately 6.4×10^6 meters, 30,000 times larger).

Density of a Neutron Star

Neutron stars form when the core of a very massive star undergoes gravitational collapse, causing the star's outer layers to explode in a supernova. Composed almost completely of neutrons, they are the densest-known stars in the universe, with densities comparable to the average density of an atomic nucleus. A neutron star in a faraway galaxy has a mass equal to 2.4 solar masses (1 solar mass = $M_\odot$ = mass of the sun = 1.99×10^{30} kg) and a diameter of 26 km.

a. What is the density of this neutron star?

b. How does this neutron star's density compare to the density of a uranium nucleus, which has a diameter of about 15 fm (1 fm = 10^{-15} m)?

Solution: We can treat both the neutron star and the U-235 nucleus as spheres. Then the density for both is given by:

$$d = \frac{m}{V} \text{ with } V = \frac{4}{3}\pi r^3$$

a. The radius of the neutron star is,

$\frac{1}{2} \times 26\,\text{km} = \frac{1}{2} \times 2.6 \times 10^4$ m = 1.3 × 10⁴ m, so the density of the neutron star is:

$$d = \frac{m}{V} = \frac{m}{\frac{4}{3}\pi r^3} \frac{2.4(1.99\text{x}10^{30}\,\text{kg})}{\frac{4}{3}\pi r^3(1.3\times10^4\,\text{m})^3} = 5.2\times10^{17}\ \text{kg/ m}^3$$

b. The radius of the U-235 nucleus is,

$\frac{1}{2} \times 15 \times 10^{-15}\ \text{m} = 7.5 \times 10^{-15}$ m, so the density of the U-235 nucleus is:

$$d = \frac{m}{V} = \frac{m}{\frac{4}{3}\pi r^3} = \frac{235\,\text{amu}\left(\frac{1.66\times10^{-27}\,\text{kg}}{1\,\text{amu}}\right)}{\frac{4}{3}\pi(7.5\times10^{-15}\,\text{m})^3} = 2.2\times10^{17}\ \text{kg/ m}^3$$

These values are fairly similar (same order of magnitude), but the nucleus is more than twice as dense as the neutron star.

To hold positively charged protons together in the very small volume of a nucleus requires very strong attractive forces because the positively charged protons repel one another strongly at such short distances. The force of attraction that holds the nucleus together is the strong nuclear force. (The strong force is one of the four fundamental forces that are known to exist. The others are the electromagnetic force, the gravitational force, and the nuclear weak force.) This force acts between protons, between neutrons, and between protons and neutrons. It is very different from the electrostatic force that holds negatively charged electrons around a positively charged nucleus (the attraction between opposite charges). Over distances less than 10^{-15} meters and within the nucleus, the strong nuclear force is much stronger than electrostatic repulsions between protons; over larger distances and outside the nucleus, it is essentially nonexistent.

Nuclear Binding Energy

As a simple example of the energy associated with the strong nuclear force, consider the helium atom composed of two protons, two neutrons, and two electrons. The total mass of these six subatomic particles may be calculated as:

$$\underbrace{(2\times1.0073\,\text{amu})}_{\text{protons}} + \underbrace{(2\times1.0087\,\text{amu})}_{\text{neutrons}} + \underbrace{(2\times0.00055\,\text{amu})}_{\text{elctrons}} = 4.0331\,\text{amu}$$

However, mass spectrometric measurements reveal that the mass of an ${}^{4}_{2}He$ atom is 4.0026 amu, less than the combined masses of its six constituent subatomic particles. This difference between the calculated and experimentally measured masses is known as the mass defect of the atom. In the case of helium, the mass defect indicates a “loss” in mass of 4.0331 amu – 4.0026 amu = 0.0305 amu. The loss in mass accompanying the formation of an atom from protons, neutrons, and electrons is due to the conversion of that mass into energy that is evolved as the atom forms. The nuclear binding energy is the energy produced when the atoms' nucleons are bound together; this is also the energy needed to break a nucleus into its constituent protons and neutrons. In comparison to chemical bond energies, nuclear binding energies are *vastly* greater. Consequently, the energy changes associated with nuclear reactions are vastly greater than are those for chemical reactions.

The conversion between mass and energy is most identifiably represented by the mass-energy equivalence equation as stated by Albert Einstein:

$$E = mc^2$$

Where E is energy, m is mass of the matter being converted, and c is the speed of light in a vacuum. This equation can be used to find the amount of energy that results when matter is converted into energy. Using this mass-energy equivalence equation, the nuclear binding energy of a nucleus may be calculated from its mass defect, as demonstrated in Example. A variety of units are commonly used for nuclear binding energies, including electron volts (eV), with 1 eV equaling the amount of energy necessary to the move the charge of an electron across an electric potential difference of 1 volt, making 1 eV = 1.602×10^{-19} J.

Calculation of Nuclear Binding Energy

Determine the binding energy for the nuclide ${}^{4}_{2}\text{He}$ in:

a. Joules per mole of nuclei

b. Joules per nucleus

c. MeV per nucleus

Solution: The mass defect for a ${}^{4}_{2}\text{He}$ nucleus is 0.0305 amu, as shown previously. Determine the binding energy in joules per nuclide using the mass-energy equivalence equation. To accommodate the requested energy units, the mass defect must be expressed in kilograms (1 J = 1 kg m²/s²).

a. First, express the mass defect in g/mol. This is easily done considering the numerical equivalence of atomic mass (amu) and molar mass (g/mol) that results from the definitions of the amu and mole units (refer to the previous discussion in the chapter on atoms, molecules, and ions if needed). The mass defect is therefore 0.0305 g/mol. To accommodate the units of the other terms in the mass-energy equation, the mass must be expressed in kg, since 1 J = 1 kg m²/s². Converting grams into kilograms yields a mass defect of 3.05×10^{-5} kg/mol. Substituting this quantity into the mass-energy equivalence equation yields:

$$E = mc^2 = \frac{3.05\times10^{-15}\,\text{kg}}{\text{mol}} \times \left(\frac{2.998\times10^{8}\,\text{m}}{\text{s}}\right)^2 = 2.74\times10^{12}\ \text{kg}\,\text{m}^2\,\text{s}^{-2}\,\text{mol}^{-1}$$

$$= 2.74\times10^{12}\ \text{J}\,\text{mol}^{-1} = 2.74\,\text{TJ}\,\text{mol}^{-1}$$

Note that this tremendous amount of energy is associated with the conversion of a very small amount of matter (about 30 mg, roughly the mass of typical drop of water).

b. The binding energy for a single nucleus is computed from the molar binding energy using Avogadro's number:

$$E = 2.74 \times 10^{12}\ \text{J mol}^{-1} \times \frac{1\ \text{mol}}{6.022 \times 10^{23}\ \text{nuclei}} = 4.55 \times 10^{-12}\ \text{J} = 4.55\ \text{pJ}$$

c. Recall that 1 eV = 1.602×10^{-19} J. Use the binding energy computed above:

$$E = 4.55 \times 10^{-12}\ \text{J} \times \frac{1eV}{1.602 \times 10^{-19}\ \text{J}} = 2.84 \times 10^{7}\ \text{eV} = 28.4\ \text{MeV}$$

Because the energy changes for breaking and forming bonds are so small compared to the energy changes for breaking or forming nuclei, the changes in mass during all ordinary chemical reactions are virtually undetectable., the most energetic chemical reactions exhibit enthalpies on the order of thousands of kJ/mol, which is equivalent to mass differences in the nanogram range (10^{-9} g). On the other hand, nuclear binding energies are typically on the order of billions of kJ/mol, corresponding to mass differences in the milligram range (10^{-3} g).

Nuclear Stability

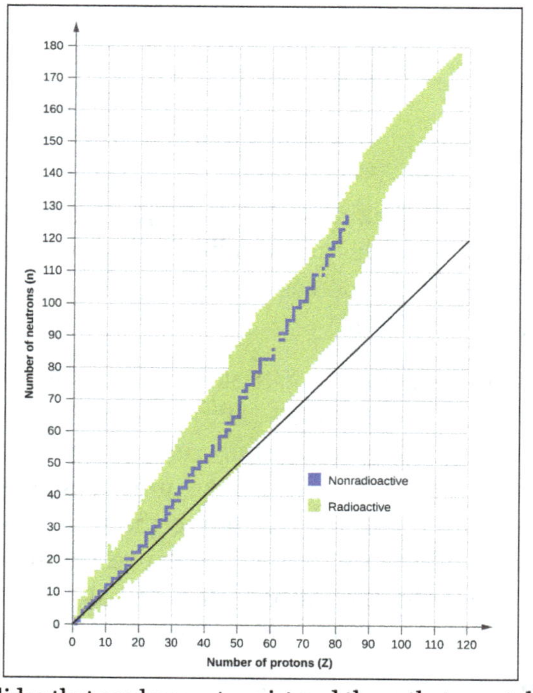

This plot shows the nuclides that are known to exist and those that are stable. The stable nuclides are indicated in blue, and the unstable nuclides are indicated in green. Note that all isotopes of elements with atomic numbers greater than 83 are unstable. The solid line is the line where n = Z.

A nucleus is stable if it cannot be transformed into another configuration without adding energy from the outside. Of the thousands of nuclides that exist, about 250 are stable. A plot of the number of neutrons versus the number of protons for stable nuclei reveals that the stable isotopes fall into a narrow band. This region is known as the band of stability (also called the belt, zone, or valley of stability). The straight line in Figure represents nuclei that have a 1:1 ratio of protons to neutrons (n:p ratio). Note that the lighter stable nuclei, in general, have equal numbers of protons and neutrons. For example, nitrogen-14 has seven protons and seven neutrons. Heavier stable nuclei, however, have increasingly more neutrons than protons. For example: iron-56 has 30 neutrons and 26 protons, an n:p ratio of 1.15, whereas the stable nuclide lead-207 has 125 neutrons and 82 protons, an n:p ratio equal to 1.52. This is because larger nuclei have more proton-proton repulsions, and require larger numbers of neutrons to provide compensating strong forces to overcome these electrostatic repulsions and hold the nucleus together.

The nuclei that are to the left or to the right of the band of stability are unstable and exhibit radioactivity. They change spontaneously (decay) into other nuclei that are either in, or closer to, the band of stability. These nuclear decay reactions convert one unstable isotope (or radioisotope) into another, more stable, isotope.

Several observations may be made regarding the relationship between the stability of a nucleus and its structure. Nuclei with even numbers of protons, neutrons, or both are more likely to be stable. Nuclei with certain numbers of nucleons, known as magic numbers, are stable against nuclear decay. These numbers of protons or neutrons (2, 8, 20, 28, 50, 82, and 126) make complete shells in the nucleus. These are similar in concept to the stable electron shells observed for the noble gases. Nuclei that have magic numbers of both protons and neutrons, such as ${}^{4}_{2}He, {}^{16}_{8}O, {}^{40}_{20}Ca$, and ${}^{208}_{82}Pb$, are called "double magic" and are particularly stable. These trends in nuclear stability may be rationalized by considering a quantum mechanical model of nuclear energy states analogous to that used to describe electronic states earlier in this textbook.

Table: Stable Nuclear Isotopes.

Number of Stable Isotopes	Proton Number	Neutron Number
157	even	even
53	even	odd
50	odd	even
5	odd	odd

The relative stability of a nucleus is correlated with its binding energy per nucleon, the total binding energy for the nucleus divided by the number or nucleons in the nucleus. For instance, the binding energy for a He nucleus is 28.4 MeV. The binding energy per nucleon for a ${}^{4}_{2}He$ nucleus is therefore:

$$\frac{28.4\,\text{MeV}}{4\,\text{nucleons}} = 7.10\,\text{MeV/ nucleon}$$

We learn how to calculate the binding energy per nucleon of a nuclide on the curve shown in figure.

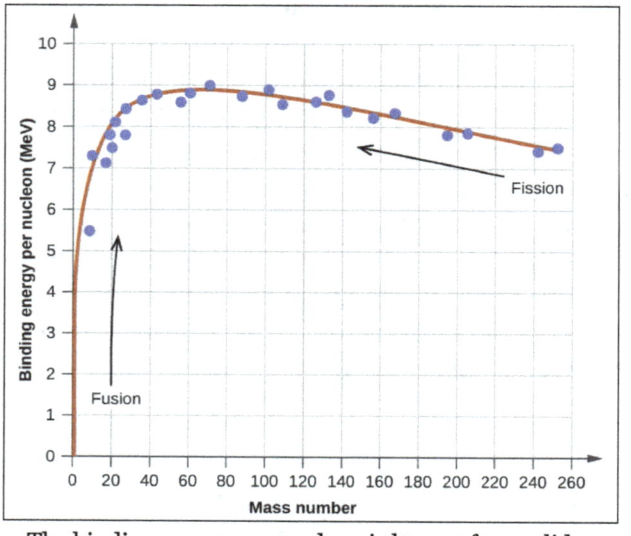

The binding energy per nucleon is largest for nuclides with mass number of approximately 56.

Nuclear Properties

The central core of an atom which contains all the atoms positive charge and most of its mass is known as an atomic nucleus. The branch of physics dealing with the study of an atomic nucleus is called nuclear physics.

Atomic nucleus was discovered by α-particle scattering experiment. In this experiment, α-particles from a radioactive source are passed through a thin gold foil of thickness 2.1×10^{-7} m. The α-particles scattered in different directions are observed through a zinc sulphide screen and a microscope. They found that most of the αparticles passed straight through the gold foil, very few α-particles scattered through a large angle and rarely a αparticle rebounds. It was concluded that an atom has the highly dense central part where almost all of the mass and all of the atom's positive charge is concentrated in a very small region as compared to that of the atom as a whole. The central part of the atom was called nucleus.

Constituents of Nucleus

All atomic nuclei except $_1H^1$ are made of the protons and neutrons. A proton has a positive charge of the same magnitude as that of an electron, but a neutron is electrically neutral. These protons and neutrons of an atom are collectively called nucleons of that atom.

A species of nucleus known as nuclide is represented by $_ZX^A$ where Z is the atomic number which indicates a number which indicates a number of protons, A is the mass number which indicates a total number of nucleons and X is the chemical symbol of the species. The number of neutrons is given by N = A – Z. For example, in $_7N^{14}$ the atomic number Z = 7, mass number A = 14, number of neutrons N = A – Z = 14 – 7 = 7 and N represents nitrogen.

Properties of Nucleus

1. Size:

The nucleus is about 10,000 times smaller than the atom. The empirical formula for the nuclear radius is $R = r_0 A^{1/3}$ where A is the mass number,

$$r_0 = 1.2 \times 10^{-15} m = 1.2 Fm.$$

2. Charge:

The nuclei consist of protons and neutrons. Protons are positively charged and neutrons are neutral. So the nuclei are positively charged. The charge of a nuclide $_ZX^A$ is Ze.

3. Mass:

Since the nucleus consists of protons and neutrons, the mass of the nucleus is the sum of masses of nucleons.

$$\text{Assumed mass} = Z\,m_p + N\,m_n$$

Where Z is the atomic number, N is neutron number m_p is a mass of a proton and m_n is a mass of a neutron.

4. Nuclear density:

The mass per unit volume of a nucleus is called nuclear density.

$$\text{Nuclear density}, \rho = \frac{\text{Nuclear mass}}{\text{Nuclear volume}}$$

$$= \frac{Am_N}{4/3\pi R^3}$$

$$= \frac{Am_N}{4/3\pi (r_0 A^{1/3})^3}$$

$$= \frac{3m_N A}{4\pi r_0^3 A}$$

Where m_N is mass of a nucleon. So,

$$\rho = \frac{3\times1.66\times10^{-27}}{4\pi(1.2\times10^{-15})^3} kg/m^3$$
$$= 2.3\times10^{17} kgm^{-3}$$

This shows that the density of nucleus is very high and is independent with A.

5. Nuclear spin and magnetic moment:

Protons and neutrons are in continuous motion in discrete quantized orbits. The nucleons have internal angular momentum called spin. As a result, they possess magnetic momentum associated with their orbital motion and spin.

Atomic Mass and Atomic Mass Unit

The abbreviated mass of atoms, nucleus and fundamental particles (electrons, protons, neutrons etc.) are called atomic mass unit. One amu is defined as 1/12 of the mass of one $_6C^{12}$ atoms. As 12 g of $_6C^{12}$ contains 6.023×10^{23} atoms of $_6C^{12}$, so mass of one $_6C^{12}$ atom is,

$$\frac{12}{6.023\times10^{23}} g$$

$$1\,\text{amu} = \frac{1}{12}\left(\frac{12}{6.023\times10^{23}}\right) g$$
$$= 1.66\times10^{-24}$$
$$\therefore 1\,\text{amu} = 1.66\times10^{-24}$$
$$= 1.66\times10^{-27} kg$$

On atomic scale, mass of protons, neutrons and electrons are,

$$m_p = 1.672622\times10^{-27} g = 1.007276\, amu$$
$$m_n = 1.674927\times10^{-27} kg = 1.008665\, amu$$
$$m_e = 9.10938\times10^{-31} kg = 0.000548580\, amu$$

Nuclear Size and Shape

The actual dimensions of the nucleus present in the atom are called nuclear size. Rutherford in the gold foil experiment showed us that the atom was mainly empty space with the nucleus at the centre and electrons revolving around it. When he fired alpha particles towards the gold foil, he noticed that 1 in 20000 particles suffered a change in direction of motion of more than 90 degrees. Rest of the 19999 article departed from their trajectory by a very small margin. This meant that the atom consisted of an empty space with most of the mass being concentrated in tiny volume in the centre.

Size of Nucleus

Through the Rutherford experiment, it was possible to obtain the size of the nucleus. By obtaining the point of closest approach of an alpha particle, we could calculate the size of the nucleus. By firing alpha particles of kinetic energy 5.5 MeV, the point of closest approach was calculated to be about 4×10^{-14}m. Since the repulsive force acting here is Coulomb repulsion, there is no contact. This means that the size of the nucleus is smaller than 4×10^{-14}m.

By firing alpha particles of greater energy and after many more iterations of the experiment, the sizes of the nuclei of various elements have been accurately measured. Through this, we have obtained a formula to measure the size of the nucleus.

$$R = R_0 A^{\frac{1}{3}}$$

Where $R_0 = 1.2\times10^{-15}$m. This means that the volume of the nucleus which is proportional to R^3 is proportional to A (mass number). One thing that should also be noticed is that there seems to be no mention of density in the equation. That's because the density of the nuclei does not vary with elements. All nuclei have the same density. The density of the nucleus is approximately 2.3×10^{17} kg.m^{-3}. This is very high compared to the density of normal things, ie water (1000 kg.m^{-3}) or air (1.225 kg.m^{-3}). This is because most of the atom is empty and all the mass is concentrated in a very tiny space.

Nuclear Density

Nuclear density is the density of the nucleus of an atom. It is the ratio of mass per unit volume inside the nucleus. Since atomic nucleus carries most of atom's mass and atomic nucleus is very small in comparison to entire atom, the nuclear density is very high.

The nuclear density for a typical nucleus can be approximately calculated from the size of the nucleus and from its mass. Typical nuclear radii are of the order 10^{-14} m. Assuming spherical shape, nuclear radii can be calculated according to following formula:

$$r = r_0 . A^{1/3}$$

where,

$$r_0 = 1.2\times10^{-15}\ \text{m} = 1.2\ \text{fm}.$$

For example, natural uranium consists primarily of isotope ^{238}U (99.28%), therefore the atomic mass of uranium element is close to the atomic mass of ^{238}U isotope (238.03u). Its radius of this nucleus will be:

$$r = r_0 . A^{1/3} = 7.44\,\text{fm}.$$

Assuming it is spherical, its volume will be:

$$V = 4\pi r^3 / 3 = 1.73\times10^{-42}\,\text{m}^3.$$

The usual definition of nuclear density gives for its density:

$$\rho_{\text{nucleus}} = m / V = 238\times1.66\times10^{-27} / (1.73\times10^{-42}) = 2.3\times10^{17}\,\text{kg} / \text{m}^3.$$

Thus, the density of nuclear material is more than 2.10^{14} times greater than that of water. It is an immense density. The descriptive term *nuclear density* is also applied to situations where similarly high densities occur, such as within neutron stars. Such immense densities are also found in neutron stars.

Density of Neutron Star

The densest material found on earth is the metal osmium, but its density pales by comparison to the densities of exotic astronomical objects such as white dwarf stars and neutron stars.

A neutron star is the collapsed core of a large star (usually of a red giant). Neutron stars are the smallest and densest stars known to exist and they are rotating extremely rapidly. A neutron star is basically a giant atomic nucleus about 11 km in diameter made especially of neutrons. It is believed that under the immense pressures of a collapsing massive stars going supernova it is possible for the electrons and protons to combine to form neutrons via electron capture, releasing a huge amount of neutrinos.

They are so dense that one teaspoon of its material would have a mass over 5.5×10^{12} kg. It is assumed they have densities of 3.7×10^{17} to 6×10^{17} kg/m^3, which is comparable to the approximate density of an atomic nucleus of 2.3×10^{17} kg/m^3.

Mass of Nuclei and Binding Energy

A nucleus consists of Z protons and N neutrons. Let the mass of the nucleus be M (Z, N).

If a nucleus is assumed to be a simple collection of Z protons and N neutrons, the mass of the nucleus would be just the sum of the masses of these constituent nucleons, i.e. $ZM_p + N M_n$ where M_p is the mass of proton and M_n is that of neutron.

However, a nucleus is not a simple collection of protons and neutrons (nucleons), but they strongly combine with each other through a strong interaction named the nuclear force.

In general, if two or more particles interact to combin together, then the total mass of the system would decrease to be less than the sum of the masses of the individual particles. The stronger the interaction becomes, the more the mass decreases. This decrease of the mass of the system is called the mass defect.

The mass defect of a nucleus of proton number Z and neutron number N is defined by,

$$\text{Mass Defect} = (Z\,M_p + NM_{\mathrm{n}}) - M(Z,N)$$

According to the Special Theory of Relativity proposed by A. Einstein, it was made clear that mass and energy are equivalent (the Mass-Energy Equivalence), which is expressed by an equality,

$$E = mc^2; \quad (c = \text{light speed} \approx 3\times10^8 \text{ m/ s})$$

On the right-hand side of this equality, mass is multiplied by the square of the speed of light. This implies that very small amounts of mass may be converted into a very large amount of energy, and vice versa.

In the World of the Atomic Nucleus, a usually treated mass is, in general, very small. Therefore mass in this world is sometimes expressed in units of energy by being converted into energy with the above-mentioned Einstein's mass-energy equivalence.

For example, the proton mass is $M_p = 1.6726\times10^{-27}$ kg which is a quite small amount. Hence, this is converted into energy by being multiplied by the square of c i.e.

$$M_p c^2 = 938.27\,\text{MeV}$$

Accordingly, the proton mass is expressed as,

$$M_p = 1.6726\times10^{-27}\text{ kg} = 938.27\,\text{MeV}/\,c^2$$

In this way, the unit of mass,

$$\text{MeV}/\,c^2$$

is often used in the World of the Atomic Nucleus.

Since a nucleus consists of A nucleons (Z protons and N neutrons; A = Z + N) being bound together, the mass defect occurs as explained at the top of the present page. The binding energy B (Z, N) is what this mass defect is converted into with the mass-energy equivalence. It is therefore given by,

$$B(Z,N) = \{(Z\,M_{\mathrm{p}} + N\,M_{\mathrm{n}}) - M(Z,N)\}c^2$$

Where M (Z, N) is the mass of a nuclide of Z protons and N neutrons. Namely, the binding energy denotes the degree how strongly the nucleons bind together, and its value can be obtained by measuring the mass of the nuclide.

Nowadays, we can obtain the precise value of the mass of an individual nuclide using the mass spectrometer for example.

Experiments to measure the nuclear binding energies of wide range of nuclides have been carried out. The resultant data are shown in the figure below. From this figure, we can see that, the binding energy per nucleon is nearly constant (= about 8 MeV) for the nuclei of A > 20, though there are some irregularities in the region of light nuclei.

The property that the binding energy per nucleon is nearly constant is called the saturation of the nuclear binding energy. This is one of the remarkable properties of the atomic nuclei.

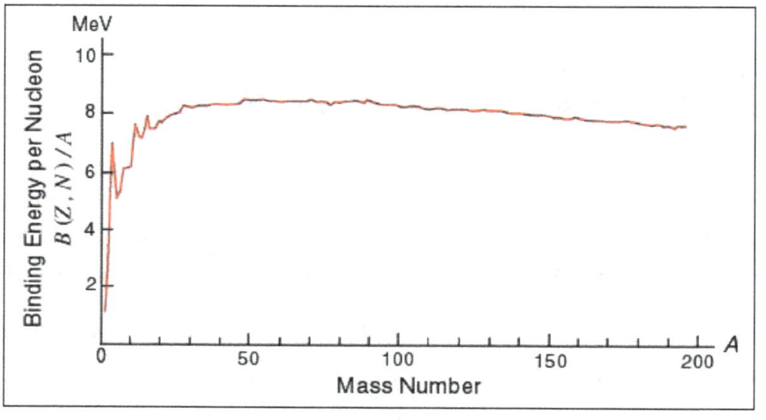

Actinide Chemistry

Actinide chemistry is critical for developing approaches for nuclear waste treatment as well as for illuminating the role of f orbitals in bonding and reactivity. A pair of studies describes new types of complexes these elements can form.

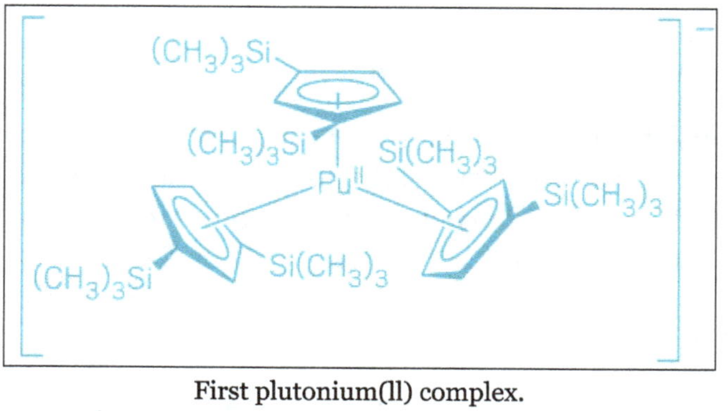

First plutonium(II) complex.

At the far end of the actinide series, the elements prefer to be divalent—nobelium(II), for example, is the only known stable state for that element in aqueous solution. Conversely, early actinides such as uranium, neptunium, and plutonium can adopt a number of oxidation states from+3 to +7 while typically avoiding the +2 state.

A team led by Matthew R. MacDonald and William J. Evans prepared the first uranium(II) complex by reducing $U^{III}[C_5H_4Si(CH_3)_3]_3$ with potassium graphite in the presence of 2.2.2-cryptand—an approach that also worked to synthesize several lanthanide(II) complexes.

It is challenging to conceive of more difficult chemistry. In addition to tackling the high reactivity of these complexes, both teams had to contend with the very real radiologic issues associated with neptunium and plutonium.

Because of radiologic safety and nuclear security concerns, the chemists had to work with small amounts of materials, synthesizing less than 100 mg of the complexes.

Andrew Gaunt, one of the leaders of the plutonium(II) complex synthesis, works in a glove box at Los Alamos National Laboratory.

Arnold and colleagues prepared several neptunium complexes in the +2, +3, and +4 oxidation states with cyclopentadienyl ligands. The complexes are generally more stable than their thorium or uranium analogs, and the researchers were able to determine that the ionic radius of neptunium shrinks by 8 picometers when it is oxidized from Np(III) to Np(IV). The product thought to be the $\{Np^{II}[C_5H_4Si(CH_3)_3]_3\}^{1-}$ complex, however, formed dark brown crystals that degraded too quickly to get X-ray diffraction data. Arnold's lab has previously synthesized another Np(II) complex with a macrocyclic ligand, trans-calixbenzenepyrrole.

Gaunt synthesized and structurally characterized the $\{Pu^{II}[C_5H_3(Si(CH_3)_3)_2]_3\}^{1-}$ complex, which forms purple crystals. The synthesis establishes plutonium as the actinide element with the largest number of confirmed oxidation states: 0 through +7, except +1.

Additional spectroscopic and theoretical analysis of the Pu(II) complex indicated that Pu(II) has a $5f^6$ ground-state electronic configuration, in contrast to d-orbital population observed for earlier actinides. The ground state for Th(II), for example, is $5f^06d^2$, whereas the ground state for U(II) is $5f^36d^1$. The results add to plutonium's reputation for unique and complex properties that represent a transition point between the chemistry of lighter and heavier actinides.

Actinides

The Actinide series contains elements with atomic numbers 89 to 103 and is the third group in the periodic table. The series is the row below the Lanthanide series, which is located underneath the main body of the periodic table. Lanthanide and Actinide Series are both referred to as Rare Earth Metals. These elements all have a high diversity in oxidation numbers and all are radioactive. The most common and known element is Uranium, which is used as nuclear fuel when its converted into plutonium, through a nuclear reaction.

The first actinides to be discovered were Uranium by Klaproth in 1789 and Thorium by Berezelius in 1829, but most of the Actinides were man-made products of the 20th century. Actinium and Protactinium are found in small portions in nature, as decay products of 253-Uranium and 238-Uranium. Microscopic amounts of Plutonium are made by neutron capture by Uranium, and yet occur naturally. Monazite is the principle Thorium ore. It is a phosphate ore that contains great amounts of Lanthanides in it. The main Uranium ore is U_3O_8 and is known as pitchblende, because it occurs in black, pitch-like masses. An example of pitchblende is located in the picture below. All elements past Uranium is man-made. Actinides require special handling, because many of them are radioactive and/or unstable. The radiation in actinides plays a large role in the chemistry and arrangement of particles in crystals.

This is a picture of U_3O_8, a uranium, pitchblende ore, by Geomartin.

Common Properties

- All are radioactive due to instability.
- Majority synthetically made by particle accelerators creating nuclear reactions and short lasting.

- All are unstable and reactive due to atomic number above 83 (nuclear stability).
- All have a silvery or silvery-white luster in metallic form.
- All have the ability to form stable complexes with ligands, such as chloride, sulfate, carbonate and acetate.
- Many of the actinides occur in nature as sea water or minerals.
- They have the ability to undergo nuclear reactions.
- The emission of radioactivity, toxicity, pyrophoricity, and nuclear criticality are properties that make them hazardous to handle.
    - Emission of Radioactivity: The types of radiation the elements possess are alpha, beta, gamma, as well as when neutrons are produced by spontaneous fissions or boron, beryllium, and fluorine react with alpha-particles.
    - Toxicity: Because of their radioactive and heavy metal characteristics, they are considered toxic elements.
    - Pyrophoricity: Many actinide metals, hydrides, carbides, alloys and other compounds may ignite at room temperature in a finely divided state, which would result from spontaneous combustion fires and spreading of radioactive contaminates.
    - Nuclear Criticality: If fissionable materials are combined, a chain reaction could occur resulting in lethal doses of radioactivity, but it depends on chemical form, isotopic composition, geometry, size of surroundings, etc.
- The interaction of Actinides when radioactive with different types of phosphors will produce pulses of light.

Table: Actinide Properties.

Atomic Number	Symbol	Name	Atomic Mass (g/mol)	Electron Configuration	Oxidation States	Year Discovered
89	Ac	Actinium	[227]	[Rn] $6d^17s^2$	+3	1899
90	Th	Thorium	232.04	[Rn] $6d^27s^2$	+2?,+3,+4	1829
91	Pa	Protactinium	231.04	[Rn] $5f^26d^17s^2$	+3,+4,+5	1913
92	U	Uranium	238.03	[Rn] $5f^36d^17s^2$	+3,+4,+5,+6	1789
93	Np	Neptunium	[237]	[Rn] $5f^46d^17s^2$	+3,+4,+5,+6,+7	1940
94	Pu	Plutonium	[244]	[Rn] $5f^67s^2$	+3,+4,+5,+6,+7	1940
95	Am	Americium	[243]	[Rn] $5f^77s^2$	+3,+4,+5,+6,+7?	1944
96	Cm	Curium	[247]	[Rn] $5f^7 6d^17s^2$	+3,+4,+5,+6?	1944
97	Bk	Berkelium	[247]	[Rn] $5f^97s^2$	+3,+4	1949
98	Cf	Californium	[251]	[Rn] $5f^{10}7s^2$	+2,+3,+4	1950

99	Es	Einsteinium	[252]	[Rn] $5f^{11}7s^2$	+2?,+3,+4?	1952
100	Fm	Fermium	[257]	[Rn] $5f^{12}7s^2$	+2,+3	1952
101	Md	Mendelevium	[258]	[Rn] $5f^{13}7s^2$	+2,+3	1955
102	No	Nobelium	[259]	[Rn] $5f^{14}7s^2$	+2,+3	1957
103	Lr	Lawrencium	[262]	[Rn] $5f^{14}6d^{1}7s^2$	+3	1961

- [#]-mass number of the longest living isotope is given when the element has no stable nucleotides.
- (?)-oxidation state is unconfirmed.

Actinides have been crucial in understanding nuclear chemistry and have provided valuable usage today, such as nuclear power. These examples illustrate their importance in understanding key concepts in nuclear chemistry and related topics.

Transition Metal

Actinides are in the f-block of the periodic table. The electron configuration of uranium is [Rn] $5f^3$ $6d^1$ $7s^2$. The reason for this arrangement unlike other conventional electron configurations such as Na with configuration of [Ne]$3s^2$. Results from difference in energy levels due to the fact that some orbitals fill in faster than others and explains why Actinides are transition metals.

The electron configurations of the actinides are due to the following:

- The energy in the 6d orbitals is lower in energy than in the 5f orbitals.
- They fill 5f orbital, 6d orbital, then 7s orbital.
- The 5f orbitals are not shielded by the filled 6s and 6p subshells.
- There is a small energy gap between the $5f^n$ $7s^2$ and $5f^{n-1}$ 6d $7s^2$ configurations.
- The 5f orbitals do not shield each other from the nucleus effectively.
- The energies of the 5f orbital drop rapidly with increasing atomic number.

Chemistry

- Actinium
    - Its chemistry is dominated by (+3) O. S.
    - Its compounds are colorless.
    - There are 29 known isotopes.
    - It does not have absorption in the UV visible region between 400-1000nm.

- ^{227}Ac is strongly radioactive and so are its decay components.
- Actinium metal is silvery solid; obtained by reduction of oxide, fluoride or chloride w/ Group 1 metals; and oxidized rapidly in moist air.
- It forms insoluble fluoride and oxalate ($Ac_2(C_2O_4)_3$*10H_2O) compounds.

- Thorium
  - It exhibits the +4 O.S. exclusively.
  - The chemistry in the +2 and +3 O.S. is restricted to iodides like ThI_2 and cyclopentadienyl $Th(C_5H_5)_3$.
  - It has wide coordination chemistry with oxygen donor ligands.
  - Thorium metal is bright and silvery-white and tarnishes to a dull black color when exposed to air. It is soft enough to be scratched with a knife and melts at 1750°. It slowly dissolves in dilute with hydrogen evolution and can be pyrophoric as a powder.
- Protactinium
  - It has been in existence longer than any other actinide.
  - ^{231}Pa has a half-life of $3.28*10^{14}$ which allows it to make chemical study easy for it.
  - It has α-emission, so it has appropriate radiochemical precautions.
  - The Pa metal is malleable, ductile, silvery, and has a melting point of about 1565°C. It is also a superconductor.
- Uranium
  - Many compounds exist between the O.S. of +3 to +6.
  - The main O.S. is +4 and +6.
  - Stability of O.S.
    - U^{3+} reduces to hydrogen.
    - U^{4+} stable in aqueous solution in the absence of air.
    - U^{5+} disproportionates rapidly into a mixture of U^{4+} and U^{6+} in aqueous solutions.
    - U^{6+} stable in aqueous solutions.
  - When pure it has a silvery appearance.

- When attacked by air, yellow film then black coating develops, it is a mix of oxide and nitride.
- Powder metal is pyrophoric in air.
- Reacts readily with hot water to prevent substances from coming into contact in nuclear reactors.

- Neptunium
  - It was the first transuranium element to be discovered in 1940.
  - There are 15 known isotopes, only ^{237}Np, w/ half-life of $2.14*10^6$ years, is useful for chemical experiments.
  - It exhibits O.S. of +3 to +7 in compounds.
  - It is a silvery metal, with a melting point of 637°C and a boiling point of 4174°C.
  - It has surface oxidation when exposed to air.
  - It is converted to NpO_2 at high temperatures.
- Plutonium
  - There are 15 known isotopes.
    - The masses range from 232 to 246.
    - The most important isotope is ^{239}Pu because it is fissionable and has a half-life of 24,100 years, which makes it easy for chemists to study.
  - It exhibits O.S. from +3 to +7.
    - The +3 and +4 O.S. are the most important, but compounds of the ions are well defined.
    - Pu^{+7} only exists under very alkaline conditions.
  - It has 6 allotropic metal forms, which makes it unusual.
    - They can form at normal pressure between room temperature and its melting point, 640°C.
    - It is dense, silvery and a reactive metal; more reactive than uranium or neptunium.
    - When attacked by air, it forms a green-gray oxide coating.
    - It reacts slowly with cold water, faster with dilute H_2SO_4, and dissolves quickly in dilute hydrochloric acid or hydrobromic acid.

- Americium
    - It has 12 known isotopes.
    - It was first made in 1944-1945 by Seaborg and his coworkers, where they decayed ^{239}Pu and ^{241}Pu to ^{241}Am, which has a half-life of 433 years.
        - ^{241}Am and ^{243}Am, which has a half-life of 7380 years are the most important isotopes, because their half-lives allow scientists to study their characteristics.
    - The metal is a slivery, ductile and very malleable.
        - It tarnishes in air slowly and dissolves in dilute hydrochloric acid quickly.
        - It reacts with heating with oxygen, halogens, and other nonmetals.
- Later Actinides (Cm, Bk, Cf, Es, Fm, Md, No, and Lr)
    - Their chemistry is of mostly the M^{+3} state.
    - They all form binary compounds, such as trihalides.
    - Curium, berkelium, and californium have the following chemistry:
        - Oxidized by air to the oxide.
        - Electropositive.
        - Reacts with hydrogen on warming to form hydrides.
        - Yields compounds on warming with group 5 and group 6 non-metals.

Solubility and Precipitation

Knowing about the solubility and precipitations of actinides, help chemists understand their properties better. Not all Actinides have the same properties when it comes solubility and precipitation, but most of the Actinides have similar traits and characteristics.

The fluorides, hydroxides and oxalates of actinides have low solubilities. The lighter trivalent, having a valence of three electrons, are unstable in aqueous solutions, such as transplutonium elements. The actinides of light weight have a higher valence; for example, oxidation states of 4, 5 and 6 where they are more stable; and they form compounds with low solubility. Trivalent Actinides can be separated from slightly acidic solutions as phosphates, mildly acidic solutions as oxalates, strongly acidic solutions as fluorides, and from basic solutions as hydroxides or hydrous oxides.

Hydrogen peroxide precipitates Pu^{+4}, Th^{+4}, Ce^{+4}, U^{+4}, Np^{+4} and Pa^{+4}. Pu^{+4}, Th^{+4}, Np^{+4}, UO^{+2} and all trivalent actinides are precipitated by carbonate-free ammonium hydroxide. The solubility of Actinide hydroxides or hydrous oxides in strong ammonium carbonate solutions allow the separation or Uranium and Thorium from other members of the Ammonium hydroxide group, such as Fe, Ti, Al and other rare earth metals. Thorium can be precipitated from ammonia with aluminum or from aluminum as a fluoride or oxalate. Uranium is precipitated with hydrogen sulfide, unless a complexing agent is present, such as Tartaric acid ($C_4H_6O_6$), and can be precipitated with Ammonium sulfide, unless Ammonium carbonate is present.

Halides

A majority of actinides form halides with halogens at specific temperatures and trihalides are the most well-known halides. The following are examples of Actinide compounds forming with halide compound creating actinide halide compounds:

$$\underbrace{Ac(OH)_3 + 3HF \xrightarrow{700^\circ C} AcF_3 + 3H_2O}_{Fluorides}$$

$$\underbrace{Ac_2O_3 + 6NH_4Cl \xrightarrow{250^\circ C} 2AcCl_3 + 6NH_3 + 3H_2O}_{chlorides}$$

$$\underbrace{Ac_2O_3 + 2AlBr_3 \xrightarrow{750^\circ C} 2AcBr_3 + Al_2O_3}_{Bromides}$$

$$\underbrace{Th + 2I_2 \xrightarrow{400^\circ C} ThI_4}_{Iodides}$$

The halides are very important binary compounds, sometimes the most important. Although they may have radioactivity that causes problems, they are useful to study to understand the trends in the Actinide series. Below are some halide characteristics of the actinide series from Actinium to Einsteinium.

- Actinium
  - Forms halides in the oxidation state of +3 only.
- Thorium
  - Forms in the 2, 3, and 4 oxidation states.
  - Exists in black and gold forms.
  - Has high conductivity.
  - Contains free electrons.
  - Has all whites are solids.
  - Forms many complexes with neutral donors.

    - Has an eight coordination number (dodecahedron, sq antiprism).
- Protactinium
    - Forms many different types of halides and several oxyhalides.
    - Chlorides ($PaCl_5$) are yellow-green, bromides ($PaBr_4$) are orange-red, fluorides (PaF_4) are brown, and iodides (PaI_3) are dark brown.
    - Have six coordinate (dimeric), eight coordinate (dodecahedral and square antiprismatic), seven coordinate, and nine coordinate.
- Uranium
    - Forms halides in oxidation states from +3 to +6.
    - Uranium (VI): UF_6 (colorless, does not require use of elemental fluorine to be made, octahedral, volatile), UCl_6 (dark green, hydroscopic, octahedral, volatile).
    - Uranium (V): rare oxidation states, unstable, six and seven coordination.
    - Uranium (IV): most important uranium halides, U^{+4} does not have reducing tendencies, UI is stable but not to hydrolysis.
    - Uranium (III): ease of oxidation of U^{+3}, all made under reducing conditions, have typical structure of actinide trihalides, UF_3 (green, 11-coordinate structure), UCl_3 and UBr_3 (red, trigonal prism), UI_3 (8-coordinate, black).

Uraninite, also known as pitchblende, is the most common ore mined to extract uranium.

- Neptunium
    - Has a range of halides in different O.S. but only fluorides are in above +4.
    - Has purple and blue-white fluorides, has green chlorides, has green and red-orange and dark red bromides and has brown iodides.

- NpF_6 forms a toxic and volatile vapor.
- There also exists oxyhalides, such as $NpOF_3$ and NpOCl.

- Plutonium
  - Has a low stability of high O.S.
  - Only consists of trihalides and tetrahalides, but PuF_6 exists though; and it is very volatile, reactive, is best kept in its gas state, does not exist in solid state, and is decomposed by its own radiation.
  - The trihalides have pastel colors: fluoride (violet-blue), chloride (blue-green), bromide (light-green), iodide (bright-green); +2 state is inaccessible.
  - Its complexes are in the +4 state and are stable except for iodides.
  - Its oxyhalides include $PuO_2Cl_2{*}H_2O$.
- Americium
  - The dihalides become insoluble for the first time.
  - They are black solids.
  - Only the fluorides have higher than +3 O.S.
  - The trihalides are important and have pastel colors: fluoride and chloride are pink, bromide is white-pale yellow, and iodide is pale yellow.
- Curium(III) Chloride
  - Is made by ^{244}Cm isotope which is converted to ^{248}Cm then heated to 1200°C and converted to oxide then chlorinated with HCl at 500°C to create $CmCl_3$.
- Californium
  - The only oxidation state s are +3 and +2 and it only forms chloride and iodide halides.
  - The following are examples of Californium halides: $CfCl_3$, CfI_3, and CfI_2.
- Einsteinium
  - Its only oxidation state is +3 and it only forms with chloride, fluorine, and even oxygen.
  - The following are examples of Einsteinium halides: Es_2O_3, $EsCl_3$, EsF_3.

Oxides

All Actinides form oxides with different oxidation states. The most common oxides are of the form M_2O_3, where M would be one of the elements in the Actinide series. The earliest actinides have a closer relation to the transition metals, where the oxidation state is equal to the number of electrons on the outer shell. The +4 state is more stable in the Actinide series than in the Lanthanides. The following are the different oxides of the Actinide elements:

- M_2O_3:Ac, Pu-Es
- MO_2:Th, U-Cf
- M_2O_5:Pa, Np
- MO_3: U, Np(?)

(?)-Neptunium is unknown in this oxide state, but scientists assume it does exist.

Oxyhalides

The oxyhalides of Actinides are not binary but some are formed by earlier Actinides, for example, an aqueous Protactinium fluoride reacts with air to make PaO_2F, Thorium can form $ThOX_2$, Neptunium can make $NpOF_3$, Plutonium forms $PuOF_4$ and UO_2F_2 is made by Uranium (III) oxide and Hydrofluoric acid.

Uranium Hydride (UH_3)

The uranium in uranium hydride has an oxidation state of +3, which is strongly reducing, so it forms a stable hydride. Uranium hydride reacts with hydrogen gas at 250 degrees and swells up into a fine black powder, which is pyrophoric in air. It decomposes at higher temperatures to hydrogen and uranium powders. It can also react with many other compounds at certain temperatures to make halides, oxides, and other compounds, including the following:

- With HBr to make UBr_3
- With H_2O to make UO_2
- With Cl_2 to make UCI_4
- With HF to make UF_4
- With H_2S to make US_2

Nuclear Fission Reaction of Uranium

$$^{235}U + {}^{1}n \rightarrow {}^{236}U \rightarrow \text{fission fragments} + \text{neutrons} + 3.20 \times 10^{-11} \text{ J.}$$

235-Uranium is bombarded with neutrons and turns into 236-Uranium. The 236-Uranium then converts into smaller pieces, 2-3 neutrons are released along with energy. These extra neutrons help create a chain reaction as more neutrons come into contact with Uranium. This uncontrolled energy eventually leads to an explosion, which is the basis of the atomic bomb.

Alpha Particles

The nucleus of Uranium emits radioactivity in the form of alpha particles. Alpha particles are Helium atoms with the atomic mass of 4 and an atomic number of 2. Alpha particles produce ions but have weak penetrating power that can easily be stopped by sheets of paper. Alpha particles produce large numbers of ions because of their collisions and near collisions with atoms, while traveling through matter. Their positive charge allows them to be deflected by electric and magnetic fields.

By using a nuclear equation and following two rules:

- The sum of mass numbers must be the same on both sides.
- The sum of atomic numbers must be the same on both sides.

We can write equations, such as the following, where ${}_{2}^{4}\text{He}$ represents the alpha particle.

$$^{238}U \rightarrow {}^{234}\text{Th} + {}_{2}^{4}\text{He}$$

The rules for writing a nuclear equation can apply to other radioactive decay processes, such as beta particle or gamma ray decay.

Beta Particles

Beta particles are deflected by the electric and magnetic fields in the opposite direction from alpha particles. Because they are not as big or massive as alpha particles, they are deflected more strongly than alpha particles are. Beta particles are electrons that came from the nuclei of atoms in nuclear decay processes. Although beta particles do not have actual atomic numbers, its charge is extremely close to the atomic number of -1. Many times a beta particle is small enough that its charge could be ignored during calculations. When we think of beta decay processes, we can look at the following equations and imagine a neutron within the nucleus of an atom converting to a proton or electron spontaneously. The proton remains in the nucleus and the electron is produced as a beta particle. The extra proton causes the atomic number to increase by one unit, but the mass number is unchanged.

- Beta particle-ejected electron from nucleus of U:

$$^{239}U \rightarrow {}^{239}Np + \beta$$

- Beta particle-ejected electron from nucleus of Np:

$$^{239}Np \rightarrow ^{239}Pu + \beta^-$$

- Beta particle-ejected electron from nucleus of U:

$$^{237}U \rightarrow ^{237}Np + \beta^-$$

- Beta particle-ejected electron from nucleus of Pa:

$$^{234}Pa \rightarrow ^{234}U + \beta^-$$

- Beta particle-ejected electron from nucleus of Th:

$$^{234}Th \rightarrow ^{234}Pa + \beta^-$$

Gamma Rays

Gamma rays (γγ) are radiation that is emitted when the nucleus is in an excited state due to excess energy. This energy is then released in the form of gamma rays. Gamma Rays are a form of electromagnetic energy such as visible light and hence are undeflected by electric and magnetic fields. The following are some examples of gamma ray reactions:

$$^{230}\mathrm{Th} + ^{1}_{0}\mathrm{n} \rightarrow ^{231}\mathrm{Th} + \gamma$$

$$^{238}\mathrm{U} + ^{1}_{0}\mathrm{n} \rightarrow ^{239}\mathrm{U} + \gamma$$

$$^{230}\mathrm{Th} \rightarrow ^{230}\mathrm{Th} + \gamma$$

Applications

- Plutonium and Uranium are used as nuclear fuels and in weaponry.
- Thorium (ThO_2) is used as an incandescent gas mantle from the 1880s. They are now used for cooking, gas lanterns, and for decorative gas lights.
- The 238-Plutonium isotope.
    - Powered the Apollo-12 Lunar Mission's left behind generator. It generated under 1.5 kW of heat converted to electricity by thermoelectric elements.
    - Was a power source for orbiting satellites and missions to take pictures of Jupiter.

  - Powers pacemakers for the heart. The nuclear energy creates a longer lifetime use of the device.

- The 241-Americium isotope is used as ionizing sources for smoke detectors.
- Uranium is used in nuclear reactors in the form of fuel in rods suspended in water under a pressure of 70 to 150 atmospheres.

References

- Nuclear-Chemistry: chemistryexplained.com, Retrieved 2 March, 2019
- Nuclear-structure-and-stability, chemistry: opentextbc.ca, Retrieved 18 July, 2019
- Size-of-the-nucleus, physics: byjus.com, Retrieved 17 May, 2019
- Density-nuclear-matter, what-is-density-physics, thermodynamic-properties, thermodynamic: nuclear-power.net, Retrieved 5 January, 2019
- Nuclear-mass: phys.kyushu-u.ac.jp, Retrieved 25 June, 2019
- Actinide-chemistry-expands-new-oxidation: acs.org, Retrieved 23 April, 2019
- Properties-and-Reactions-of-The-Actinides, Inorganic-Chemistry: libretexts.org, Retrieved 28 February, 2019

2

Radiation Chemistry

The branch of nuclear chemistry which deals with the study of chemical effects of radiation on matter is known as radiation chemistry. The techniques used in radiation chemistry include exposure to a steady radiation source and exposure to a pulsed radiation source. The topics elaborated in this chapter will help in gaining a better perspective about radiation chemistry and these techniques used within it.

Nuclear Radiation

Nuclear radiation refers to the particles and photons emitted during reactions that involve the nucleus of an atom. During the fission of U-235 the nuclear radiation that is released contains neutrons and gamma ray photons.

Radiation is energy in the process of being transmitted. It may take such forms as light, or tiny particles much too small to see. Visible light, the ultra-violet light we receive from the sun, and transmission signals for TV and radio communications are all forms of radiation that are common in our daily lives. These are all generally referred to as 'non-ionizing' radiation, though at least some ultra-violet radiation is considered to be ionizing.

Radiation particularly associated with nuclear medicine and the use of nuclear energy, along with X-rays, is 'ionizing' radiation, which means that the radiation has sufficient energy to interact with matter, especially the human body, and produce ions, *i.e.* it can eject an electron from an atom.

X-rays from a high-voltage discharge were discovered in 1895, and radioactivity from the decay of particular isotopes was discovered in 1896. Many scientists then undertook study of these, and especially their medical applications. This led to the identification of different kinds of radiation from the decay of atomic nuclei, and understanding of the nature of the atom. Neutrons were identified in 1932, and in 1939 atomic fission was discovered by irradiating uranium with neutrons. This led on to harnessing the energy released by fission.

A United Nations Environment Programme (UNEP) stated: "Today, we know more about the sources and effects of exposure to ionizing radiation than to almost any other hazardous agent, and the scientific community is constantly updating and analysing its knowledge. The sources of radiation causing the greatest exposure of the general public are not necessarily those that attract the most attention. In fact, the greatest exposure is caused by natural sources ever present in the environment, and the major contributor to exposure from artificial sources is the use of radiation in medicine worldwide."

Types of Radiation

Nuclear radiation arises from hundreds of different kinds of unstable atoms. While many exist in nature, the majority are created in nuclear reactions. Ionizing radiation which can damage living tissue is emitted as the unstable atoms (radionuclides) change ('decay') spontaneously to become different kinds of atoms.

The principal kinds of ionizing radiation are:

Alpha Particles

These are helium nuclei consisting of two protons and two neutrons and are emitted from naturally-occurring heavy elements such as uranium and radium, as well as from some man-made transuranic elements. They are intensely ionizing but cannot penetrate the skin, so are dangerous only if emitted inside the body.

Beta Particles

These are fast-moving electrons emitted by many radioactive elements. They are more penetrating than alpha particles, but easily shielded – the most energetic of them can be stopped by a few millimetres of wood or aluminium. They can penetrate a little way into human flesh but are generally less dangerous to people than gamma radiation. Exposure produces an effect like sunburn, but which is slower to heal. The weakest of them, such as from tritium, are stopped by skin or cellophane. Beta-radioactive substances are also safe if kept in appropriate sealed containers.

Gamma Rays

These are high-energy electromagnetic waves much the same as X-rays. They are emitted in many radioactive decays and may be very penetrating, so require more substantial shielding. Gamma rays are the main hazard to people dealing with sealed radioactive materials used, for example, in industrial gauges and radiotherapy machines. Radiation dose badges are worn by workers in exposed situations to detect them and hence monitor exposure. All of us receive about 0.5-1 mSv per year of gamma radiation from rocks, and in some places, much more. Gamma activity in a substance (*e.g.* rock) can be measured with a scintillometer or Geiger counter.

X-rays are also electromagnetic waves and ionizing, virtually identical to gamma rays, but not nuclear in origin. They are produced in a vacuum tube where an electron beam from a cathode is fired at target material comprising an anode, so are produced on demand rather than by inexorable physical processes. (However the effect of this radiation does not depend on its origin but on its energy.)

Cosmic radiation consists of very energetic particles, mostly high-energy protons, which bombard the Earth from outer space. They comprise about one-tenth of natural background exposure at sea level, and more at high altitudes.

Neutrons are uncharged particles mostly released by nuclear fission (the splitting of atoms in a nuclear reactor), and hence are seldom encountered outside the core of a nuclear reactor. Thus they are not normally a problem outside nuclear plants. Fast neutrons can be very destructive to human tissue. Neutrons are the only type of radiation which can make other, non-radioactive materials, become radioactive.

Large nuclei can fission spontaneously, since the so-called strong nuclear force holding each nucleus together is not overwhelmingly stronger than the repulsive force of charged protons.

Units of Radiation and Radioactivity

In order to quantify how much radiation we are exposed to in our daily lives and to assess potential health impacts as a result, it is necessary to establish a unit of measurement. The basic unit of radiation dose absorbed in tissue is the gray (Gy), where one gray represents the deposition of one joule of energy per kilogram of tissue.

However, since neutrons and alpha particles cause more damage per gray than gamma or beta radiation, another unit, the sievert (Sv) is used in setting radiological protection standards. This weighted unit of measurement takes into account biological effects of different types of radiation and indicates the equivelent dose. One gray of beta or gamma radiation has one sievert of biological effect, one gray of alpha particles has 20 Sv effect and one gray of neutrons is equivalent to around 10 Sv (depending on their energy). Since the sievert is a relatively large value, dose to humans is normally measured in millisieverts (mSv), one-thousandth of a sievert.

Note that Sv and Gy measurements are accumulated over time, whereas damage (or effect) depends on the actual dose rate, *e.g.* mSv per day or year, Gy per day in radiotherapy.

The becquerel (Bq) is a unit or measure of actual radioactivity in material (as distinct from the radiation it emits, or the human dose from that), with reference to the number of nuclear disintegrations per second (1 Bq = 1 disintegration/sec). Quantities of radioactive material are commonly estimated by measuring the amount of intrinsic radioactivity in becquerels – one Bq of radioactive material is that amount which has an average of one disintegration per second, *i.e.* an activity of 1 Bq. This may be spread through a very large mass.

Table: Radioactivity of some natural and other materials.

1 adult human (65 Bq/kg)	4500 Bq
1 kg of coffee	1000 Bq
1 kg of brazil nuts	400 Bq
1 banana	15 Bq
The air in a 100 sq metre Australian home (radon)	3000 Bq
The air in many 100 sq metre European homes (radon)	Up to 30 000 Bq
1 household smoke detector (with americium)	30 000 Bq
Radioisotope for medical diagnosis	70 million Bq
Radioisotope source for medical therapy	100 000 000 million Bq (100 TBq)
1 kg 50-year old vitrified high-level nuclear waste	10 000 000 million Bq (10 TBq)
1 luminous Exit sign (1970s)	1 000 000 million Bq (1 TBq)
1 kg uranium	25 million Bq
1 kg uranium ore (Canadian, 15%)	25 million Bq
1 kg uranium ore (Australian, 0.3%)	500 000 Bq
1 kg low level radioactive waste	1 million Bq
1 kg of coal ash	2000 Bq
1 kg of granite	1000 Bq
1 kg of superphosphate fertilizer	5000 Bq

Though the intrinsic radioactivity is the same, the radiation dose received by someone handling a kilogram of high-grade uranium ore will be much greater than for the same exposure to a kilogram of separated uranium, since the ore contains a number of short-lived decay products, while the uranium has a very long half-life.

Older units of radiation measurement continue in use in some literature:

- 1 gray = 100 rads
- 1 sievert = 100 rem
- 1 becquerel = 27 picocuries or 2.7×10^{-11} curies

One curie was originally the activity of one gram of radium-226, and represents 3.7×10^{10} disintegrations per second (Bq).

The Working Level Month (WLM) has been used as a measure of dose for exposure to radon and in particular, radon decay products.

Since there is radioactivity in many foodstuffs, there has been a whimsical suggestion that the Banana Equivalent Dose from eating one banana be adopted for popular reference. This is about 0.0001 mSv.

Routine Sources of Radiation

Radiation can arise from human activities or from natural sources. Most radiation exposure is from natural sources. These include: radioactivity in rocks and soil of the

Earth's crust; radon, a radioactive gas given out by many volcanic rocks and uranium ore; and cosmic radiation. The human environment has always been radioactive and accounts for up to 85% of the annual human radiation dose.

Radiation arising from human activities typically accounts for up to 20% of the public's exposure every year as global average. In the USA by 2006 it averaged about half of the total. This radiation is no different from natural radiation except that it can be controlled. X-rays and other medical procedures account for most exposure from this quarter. Less than 1% of exposure is due to the fallout from past testing of nuclear weapons or the generation of electricity in nuclear, as well as coal and geothermal, power plants.

Backscatter X-ray scanners being introduced for airport security will gives exposure of up to 5 microsieverts (μSv), compared with 5 μSv on a short flight and 30 μSv on a long intercontinental flight across the equator, or more at higher latitudes – by a factor of 2 or 3. Aircrew can receive up to about 5 mSv/yr from their hours in the air, while frequent flyers can score a similar increment. On average, nuclear power workers receive a lower annual radiation dose than flight crew, and frequent flyers in 250 hours would receive 1 mSv.

The maximum annual dose allowed for radiation workers is 20 mSv/yr, though in practice, doses are usually kept well below this level. In comparison, the average dose received by the public from nuclear power is 0.0002 mSv/yr, which is of the order of 10,000 times smaller than the total yearly dose received by the public from background radiation.

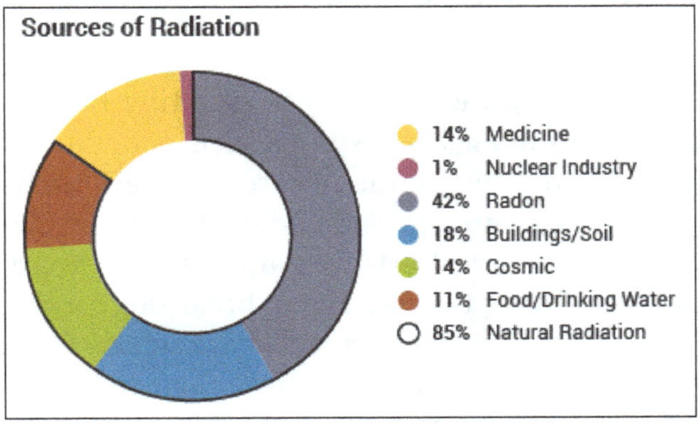

Natural Background Radiation, Radon

Naturally occurring background radiation is the main source of exposure for most people, and provides some perspective on radiation exposure from nuclear energy. Much of it comes from primordial radionuclides in the Earth's crust, and materials from it. Potassium-40, uranium-238 and thorium-232 with their decay products are the main source.

The average dose received by all of us from background radiation is around 2.4 mSv/ yr, which can vary depending on the geology and altitude where people live – ranging between 1 and 10 mSv/yr, but can be more than 50 mSv/yr. The highest known level of background radiation affecting a substantial population is in Kerala and Madras states in India where some 140,000 people receive doses which average over 15 millisievert per year from gamma radiation, in addition to a similar dose from radon. Comparable levels occur in Brazil and Sudan, with average exposures up to about 40 mSv/yr to many people. (The highest level of natural background radiation recorded is on a Brazilian beach: 800 mSv/yr, but people don't live there.)

Several places are known in Iran, India and Europe where natural background radiation gives an annual dose of more than 100 mSv to people and up to 260 mSv (at Ramsar in Iran, where some 200,000 people are exposed to more than 10 mSv/yr). Lifetime doses from natural radiation range up to several thousand millisievert. However, there is no evidence of increased cancers or other health problems arising from these high natural levels. The millions of nuclear workers that have been monitored closely for 50 years have no higher cancer mortality than the general population but have had up to ten times the average dose. People living in Colorado and Wyoming have twice the annual dose as those in Los Angeles, but have lower cancer rates. Misasa hot springs in western Honshu, a Japan Heritage site, attracts people due to having high levels of radium (up to 550 Bq/L), with health effects long claimed, and in a 1992 study the local residents' cancer death rate was half the Japan average. A study on 3000 residents living in an area with 60 Bq/m^3 radon (about ten times normal average) showed no health difference. Hot springs in China have levels reaching 3270 Bq/L radon-222 (Liaoning sanatorium), 2720 Bq/L (Tanghe hot spring) and 230 Bq/L (Puxzhe hot spring), though associated exposure from airborne radon are low.

Radon is a naturally occurring radioactive gas resulting from the decay of uranium-238, which concentrates in enclosed spaces such as buildings and underground mines, particularly in early uranium mines where it sometimes became a significant hazard before the problem was understood and controlled by increased ventilation. Radon has decay products that are short-lived alpha emitters and deposit on surfaces in the respiratory tract during the passage of breathing air. At high radon levels, this can cause an increased risk of lung cancer, particularly for smokers. (Smoking itself has a very much greater lung cancer effect than radon.) People everywhere are typically exposed to around 0.2 mSv/yr, and often up to 3 mSv/yr, due to radon (mainly from inhalation in their homes) without apparent ill-effect. Where deemed necessary, radon levels in buildings and mines can be controlled by ventilation, and measures can be taken in new constructions to prevent radon from entering buildings.

However, radon levels of up to 3700 Bq/m^3 in some dwellings at Ramsar in Iran have no evident ill-effect. Here, a study showed that the highest lung cancer mortality rate was where radon levels were normal, and the lowest rate was where radon concentrations in dwellings were highest. The ICRP recommends keeping workplace radon levels below 300 Bq/m³, equivalent to about 10 mSv/yr. Above this, workers should be considered as occupationally exposed, and subject to the same monitoring as nuclear industry workers. The normal indoor radon concentration ranges from 10 to 100 Bq/m³, but may naturally reach 10,000 Bq/m³, according to UNEP.

Table: Public exposure to natural radiation.

Source of exposure		Annual effective dose (mSv)	
		Average	Typical range
Cosmic radiation	Directly ionizing and photon component	0.28	
	Neutron component	0.10	
	Cosmogenic radionuclides	0.01	
Total cosmic and cosmogenic		0.39	0.3–1.0
External terrestrial radiation	Outdoors	0.07	
	Indoors	0.41	
Total external terrestrial radiation		0.48	0.3-1.0
Inhalation	Uranium and thorium series	0.006	
	Radon (Rn-222)	1.15	
	Thoron (Rn-220)	0.1	
Total inhalation exposure		1.26	0.2-10
Ingestion	K-40	0.17	
	Uranium and thorium series	0.12	
Total ingestion exposure		0.29	0.2-1.0
Total		2.4	1.0-13

US Naval Reactors' average annual occupational exposure was 0.06 mSv per person in 2013, and no personnel have exceeded 20 mSv in any year in the 34 years to then. The average occupational exposure of each person monitored at Naval Reactors' facilities since 1958 is 1.03 mSv per year.

Effects of Ionizing Radiation

Some of the ultraviolet (UV) radiation from the sun is considered ionizing radiation, and provides a starting point in considering its effects. Sunlight UV is important in producing vitamin D in humans, but too much exposure produces sunburn and, potentially, skin cancer. Skin tissue is damaged, and that damage to DNA may not be repaired properly, so that over time, cancer develops and may be fatal. Adaptation from repeated low exposure can decrease vulnerability. But exposure to sunlight is quite properly sought after in moderation, and not widely feared.

Our knowledge of the effects of shorter-wavelength ionizing radiation from atomic nuclei derives primarily from groups of people who have received high doses. The main difference from UV radiation is that beta, gamma and X-rays can penetrate the skin. The risk associated with large doses of this ionizing radiation is relatively well established. However, the effects, and any risks associated with doses under about 200 mSv, are less obvious because of the large underlying incidence of cancer caused by other factors. Benefits of lower doses have long been recognised, though radiation protection standards assume that any dose of radiation, no matter how small, involves a possible risk to human health. However, available scientific evidence does not indicate any cancer risk or immediate effects at doses below 100 mSv per year. At low levels of exposure, the body's natural mechanisms usually repair radiation damage to DNA in cells soon after it occurs. However, high-level irradiation overwhelms those repair mechanisms and is harmful. Dose rate is as important as overall dose.

The UN Scientific Commission on the Effects of Atomic Radiation (UNSCEAR) currently uses the term low dose to mean absorbed levels below 100 mGy but greater than 10 mGy, and the term very low dose for any levels below 10 mGy. High absorbed dose is defined as more than about 1000 mGy. For beta and gamma radiation, these figures can be taken as mSv equivalent dose.

Some comparative whole-body radiation doses and their effects	
2.4 mSv/yr	Typical background radiation experienced by everyone (average 1.5 mSv in Australia, 3 mSv in North America).
1.5 to 2.5 mSv/yr	Average dose to Australian uranium miners and US nuclear industry workers, above background and medical.
Up to 5 mSv/yr	Typical incremental dose for aircrew in middle latitudes.
9 mSv/yr	Exposure by airline crew flying the New York – Tokyo polar route.
10 mSv/yr	Maximum actual dose to Australian uranium miners.
10 mSv	Effective dose from abdomen & pelvis CT scan.
20 mSv/yr	Current limit (averaged) for nuclear industry employees and uranium miners in most countries. (In Japan: 5 mSv per three months for women)
50 mSv/yr	Former routine limit for nuclear industry employees, now maximum allowable for a single year in most countries (average to be 20 mSv/yr max). It is also the dose rate which arises from natural background levels in several places in Iran, India and Europe.
50 mSv	Allowable short-term dose for emergency workers (IAEA).
100 mSv	Lowest annual level at which increase in cancer risk is evident (UNSCEAR). Above this, the probability of cancer occurrence (rather than the severity) is assumed to increase with dose. No harm has been demonstrated below this dose. Allowable short-term dose for emergency workers taking vital remedial actions (IAEA). Dose from four months on international space station orbiting 350 km up.
130 mSv/yr	Long-term safe level for public after radiological incident, measured 1 m above contaminated ground, calculated from published hourly rate x 0.6. Risk too low to justify any action below this (IAEA).
170 mSv/wk	7-day provisionally safe level for public after radiological incident, measured 1 m above contaminated ground (IAEA).
250 mSv	Allowable short-term dose for workers controlling the 2011 Fukushima accident, set as emergency limit elsewhere.

250 mSv/yr	Natural background level at Ramsar in Iran, with no identified health effects (Some exposures reach 700 mSv/yr). Maximum allowable annual dose in emergency situations in Japan (NRA).
350 mSv/lifetime	Criterion for relocating people after Chernobyl accident.
500 mSv	Allowable short-term dose for emergency workers taking life-saving actions (IAEA).
680 mSv/yr	Tolerance dose level allowable to 1955 (assuming gamma, X-ray and beta radiation).
700 mSv/yr	Suggested threshold for maintaining evacuation after nuclear accident. (IAEA has 880 mSv/yr over one month as provisionally safe.
800 mSv/yr	Highest level of natural background radiation recorded, on a Brazilian beach.
1,000 mSv short-term	Assumed to be likely to cause a fatal cancer many years later in about 5 of every 100 persons exposed to it (*i.e.* if the normal incidence of fatal cancer were 25%, this dose would increase it to 30%). Highest reference level recommended by ICRP for rescue workers in emergency situation.
1,000 mSv short-term	Threshold for causing (temporary) radiation sickness (Acute Radiation Syndrome) such as nausea and decreased white blood cell count, but not death. Above this, severity of illness increases with dose.
5,000 mSv short-term	Would kill about half those receiving it as whole body dose within a month. (However, this is only twice a typical daily therapeutic dose applied to a very small area of the body over 4 to 6 weeks or so to kill malignant cells in cancer treatment.)
10,000 mSv short-term	Fatal within a few weeks.

The main expert body on radiation effects is the UN Scientific Commission on the Effects of Atomic Radiation (UNSCEAR), set up in 1955 and reporting to the UN General Assembly. It involves scientists from over 20 countries and publishes its findings in major reports. The UNSCEAR 2006 report dealt broadly with the Effects of Ionizing Radiation. Another valuable report, titled Low-level Radiation and its Implications for Fukushima Recovery.

In 2012 UNSCEAR reported to the UN General Assembly on radiation effects. It had been asked in 2007 "to clarify further the assessment of potential harm owing to chronic low-level exposures among large populations and also the attributability of health effects" to radiation exposure. It said that while some effects from high acute doses were clear, others including hereditary effects in human populations were not, and could not be attributed to exposure, and that this was especially true at low levels. "In general, increases in the incidence of health effects in populations cannot be attributed reliably to chronic exposure to radiation at levels that are typical of the global average background levels of radiation." Furthermore, multiplying very low doses by large numbers of individuals does not give a meaningful result regarding health effects. UNSCEAR also addressed uncertainties in risk estimation relating to cancer, particularly the extrapolations from high-dose to low-dose exposures and from acute to chronic and fractionated exposures. Earlier UNSCEAR data for leukaemia incidence among Hiroshima survivors suggested a threshold of about 400 mSv for harmful effects.

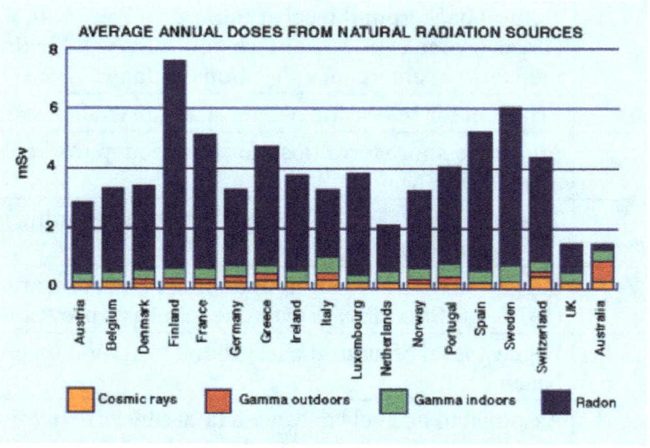

Epidemiological studies continue on the survivors of the atomic bombing of Hiroshima and Nagasaki, involving some 76,000 people exposed at levels ranging up to more than 5,000 mSv. These have shown that radiation is the likely cause of several hundred deaths from cancer, in addition to the normal incidence found in any population. From this data the International Commission on Radiological Protection (ICRP) and others estimate the fatal cancer risk as 5% per sievert exposure for a population of all ages – so one person in 100 exposed to 200 mSv could be expected to develop a fatal cancer some years later. In Western countries, about a quarter of people die from cancers, with smoking, dietary factors, genetic factors and strong sunlight being among the main causes. About 40% of people are expected to develop cancer during their lifetime even in the absence of radiation exposure beyond normal background levels. Radiation is a weak carcinogen, but undue exposure can certainly increase health risks.

In 1990, the US National Cancer Institute (NCI) found no evidence of any increase in cancer mortality among people living near to 62 major nuclear facilities. The NCI study was the broadest of its kind ever conducted and supported similar studies conducted elsewhere in the USA as well as in Canada and Europe.

About 60 years ago it was discovered that ionizing radiation could induce genetic mutations in fruit flies. Intensive study since then has shown that radiation can similarly induce mutations in plants and test animals. However there is no evidence of inherited genetic damage to humans from radiation, even as a result of the large doses received by atomic bomb survivors in Japan.

In a plant or animal cell the material (DNA) which carries genetic information necessary to cell development, maintenance and division is the critical target for radiation. Much of the damage to DNA is repairable, but in a small proportion of cells the DNA is permanently altered. This may result in death of the cell or development of a cancer, or in the case of cells forming gonad tissue, alterations which continue as genetic changes in subsequent generations. Most such mutational changes are deleterious; very few can be expected to result in improvements.

The relatively low levels of radiation allowed for members of the public and for workers in the nuclear industry are such that any increase in genetic effects due to nuclear power will be imperceptible and almost certainly non-existent. Radiation exposure levels are set so as to prevent tissue damage and minimize the risk of cancer. Experimental evidence indicates that cancers are more likely than inherited genetic damage.

Some 75,000 children born of parents who survived high radiation doses at Hiroshima and Nagasaki in 1945 have been the subject of intensive examination. This study confirms that no increase in genetic abnormalities in human populations is likely as a result of even quite high doses of radiation. Similarly, no genetic effects are evident as a result of the Chernobyl accident.

Life on Earth commenced and developed when the environment was certainly subject to several times as much radioactivity as it is now, so radiation is not a new phenomenon. If there is no dramatic increase in people's general radiation exposure, there is no evidence that health or genetic effects from radiation could ever become significant.

Low-level Radiation Effects

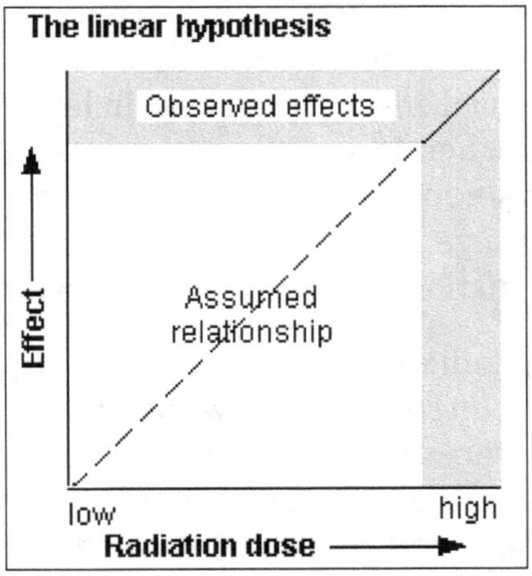

Much research has been undertaken on the effects of low-level radiation. The findings have failed to support the so-called linear no-threshold (LNT) hypothesis. This theory assumes that the demonstrated relationships between radiation dose and adverse effects at high levels of exposure also applies to low levels and provides the (deliberately conservative) basis of occupational health and other radiation protection standards.

The ICRP recommends that the LNT model should be assumed for the purpose of optimising radiation protection practices, but that it should not be used for estimating the health effects of exposures to small radiation doses received by large numbers of people over long periods of time. At low levels of exposure, the body's natural mechanism repairs radiation and other damage to cells soon after it occurs, and some adaptive response is stimulated which protects cells and tissues, as with exposure to other external agents at low levels.

A November 2009 technical report from the Electric Power Research Institute in USA drew upon more than 200 peer-reviewed publications on effects of low-level radiation and concluded that the effects of low dose-rate radiation are different and that "the risks due to those effects may be over-estimated" by the linear hypothesis. "From an epidemiological perspective, individual radiation doses of less than 100 mSv in a single exposure are too small to allow detection of any statistically significant excess cancers in the presence of naturally occurring cancers. The doses received by nuclear power plant workers fall into this category because exposure is accumulated over many years, with an average annual dose about 100 times less than 100 mSv". It quoted the US Nuclear Regulatory Commission that "since 1983, the US nuclear industry has monitored more than 100,000 radiation workers each year, and no workers have been exposed to more than 50 mSv in a year since 1989." A 2012 Massachusetts Institute of Technology study exposing mice to low-dose rate radiation for an extended period showed no signs of DNA damage, though a control group receiving the same dose acutely did show damage. This test on live animals confirms other work and epidemiological studies suggesting that people exposed to as much as 1000 mSv/yr at low dose rate will not suffer adverse health effects.

In addition, there is evidence of beneficial effect from low-level radiation (up to about 10 mSv/yr). This 'radiation hormesis' may be due to an adaptive response by the body's cells, the same as that with other toxins at low doses. In the case of carcinogens such as ionizing radiation, the beneficial effect is seen both in lower incidence of cancer and in resistance to the effects of higher doses. This potential hormetic effect is most clearly evident in the data for over 50,000 survivors of the Hiroshima bomb 1.5 to 3 km from the hypocenter, with dose range 1 to 100 mSv, compared with a large control group. Meanwhile standards for radiation exposure continue to be deliberately conservative.

In the USA, The Low-Dose Radiation Research Act of 2015 calls for an assessment of the current status of US and international low-dose radiation research. It also directs the National Academy of Sciences to "formulate overall scientific goals for the future of low-dose radiation research in the United States" and to develop a long-term research agenda to address those goals. The Act arises from a letter from a group of health physicists who pointed out that the limited understanding of low-dose health risks impairs the nation's decision-making capabilities, whether in responding to radiological events involving large populations such as the 2011 Fukushima accident or in areas such as the rapid increase in radiation-based medical procedures, the cleanup of radioactive contamination from legacy sites and the expansion of civilian nuclear energy.

Fear of Radiation Effects

The main effect of low-level radiation arises from fear, not the radiation itself. People who are conditioned to fear any level of ionizing radiation tend to take action to avoid it, and those actions are sometimes much more harmful than any exposure to low-dose radiation could be. Concerns about low doses of radiation from CT scans and X-rays are not only misguided, but may lead to suffering and deaths from avoided or delayed diagnosis. Also therapeutic benefits of nuclear medicine greatly outweigh any harm that might come from the controlled radiation exposure involved.

After the Chernobyl accident, some pregnant women in Europe sought abortions without any medical justification, the exposure levels being vastly below those likely to have any effects. Sometimes the fear is promoted by misguided governments, as in Japan where maintaining the evacuation of many people beyond a few weeks has resulted in over 1000 deaths, though exposure levels if people had returned to homes would not be hazardous except possibly in some limited areas, easily defined.

Limiting Exposure

Public dose limits for exposure from uranium mining or nuclear plants are usually set at 1 mSv/yr above background.

In most countries the current maximum permissible dose to radiation workers is 20 mSv per year averaged over five years, with a maximum of 50 mSv in any one year. This is over and above background exposure, and excludes medical exposure. The value originates from the International Commission on Radiological Protection (ICRP), and is coupled with the requirement to keep exposure as low as reasonably achievable (ALARA) – taking into account social and economic factors.

Radiation protection at uranium mining operations and in the rest of the nuclear fuel cycle is tightly regulated, and levels of exposure are monitored.

There are four ways in which people are protected from identified radiation sources:

- Limiting time. In occupational situations, dose is reduced by limiting exposure time.
- Distance. The intensity of radiation decreases with distance from its source.
- Shielding. Barriers of lead, concrete or water give good protection from high levels of penetrating radiation such as gamma rays. Intensely radioactive materials are therefore often stored or handled under water, or by remote control in rooms constructed of thick concrete or lined with lead.
- Containment. Highly radioactive materials are confined and kept out of the workplace and environment. Nuclear reactors operate within closed systems with multiple barriers which keep the radioactive materials contained.

While the release of radon in underground uranium mines makes a substantial contribution to occupational exposure on the part of the nuclear industry, the annual average effective dose to a worker in the nuclear industry overall has decreased from 4.4 mSv in the 1970s to about 1 mSv today. However, the annual average effective dose to a coal miner is still about 2.4 mSv and for other miners about 3 mSv. The mining figures are probably for underground situations.

About 23 million workers worldwide are monitored for radiation exposure, and about 10 million of these are exposed to artificial sources, mostly in the medical sector where the annual dose averages 0.5 mSv.

Standards and Regulation of Radiation Exposure

Radiation protection standards are based on the conservative assumption that the risk is directly proportional to the dose, even at the lowest levels, though there is no actual evidence of harm at low levels, below about 100 mSv as short-term dose. To the extent that cell damage is made good within a month, chronic dose rates up to 100 mSv per month could also be safe, but the standard assumption, called the 'linear no-threshold (LNT) hypothesis', discounts the contribution of any such thresholds and is recommended for practical radiation protection purposes only, such as setting allowable levels of radiation exposure of individuals.

LNT was first accepted by the International Commission on Radiological Protection (ICRP) in 1955, when scientific knowledge of radiation effects was less, and then in 1959 by the United Nations Scientific Committee on the Effects of Atomic Radiation (UNSCEAR) as a philosophical basis for radiological protection at low doses, stating outright that "Linearity has been assumed primarily for purposes of simplicity, and there may or may not be a threshold dose". (Above 100 mSv acute dose there is some scientific evidence for linearity in dose-effect.) From 1934 to 1955 a tolerance dose limit of 680 mSv/yr was recommended by the ICRP, and no evidence of harm from this – either cancer or genetic – had been documented.

The LNT hypothesis cannot properly be used for predicting the consequences of an actual exposure to low levels of radiation and it has no proper role in low-dose risk assessment. For example, LNT suggests that, if the dose is halved from a high level where effects have been observed, there will be half the effect, and so on. This would be very misleading if applied to a large group of people exposed to trivial levels of radiation and even at levels higher than trivial it could lead to inappropriate actions to avert the doses. At Fukushima following the March 2011 accident, maintaining the evacuation beyond a few days did in fact lead to about 1100 deaths, according to the Japan Reconstruction Agency.

Much of the evidence which has led to today's standards derives from the atomic bomb survivors in 1945, who were exposed to high doses incurred in a very short time. In setting occupational risk estimates, some allowance has been made for the body's ability to repair damage from small exposures, but for low-level radiation exposure the degree of protection from applying LNT may be misleading. At low levels of radiation exposure the dose-response relationship is unclear due to background radiation levels and natural incidence of cancer. However, the Hiroshima survivor data published in 1958 by UNSCEAR for leukaemia actually shows a reduction in incidence by a factor of three in the dose range 1 to 100 mSv. The threshold for increased risk here is about 400 mSv.

This is very significant in relation to concerns about radiation exposure from contaminated areas after the Chernobyl and Fukushima accidents.

The International Commission on Radiological Protection (ICRP), set up in 1928, is a body of scientific experts and a respected source of guidance on radiation protection, though it is independent and not accountable to governments or the UN. Its recommendations are widely followed by national health authorities, the EU and the IAEA. It retains the LNT hypothesis as a guiding principle.

The International Atomic Energy Agency (IAEA) is the only UN body with specific statutory responsibilities for radiation protection and safety. Its Safety Fundamentals are applied in basic safety standards and consequent Regulations. However, the UN Scientific Committee on the Effects of Atomic Radiation (UNSCEAR) set up in 1955 is the most authoritative source of information on ionizing radiation and its effects.

In any country, radiation protection standards are set by government authorities, generally in line with recommendations by the ICRP, and coupled with the requirement to keep exposure as low as reasonably achievable (ALARA) – taking into account social and economic factors. The authority of the ICRP comes from the scientific standing of its members and the merit of its recommendations.

The three key points of the ICRP's recommendations are:

- Justification. No practice should be adopted unless its introduction produces a positive net benefit.
- Optimisation. All exposures should be kept as low as reasonably achievable, economic and social factors being taken into account.
- Limitation. The exposure of individuals should not exceed the limits recommended for the appropriate circumstances.

National radiation protection standards are framed for both Occupational and Public exposure categories.

The ICRP recommends that the maximum permissible dose for occupational exposure should be 20 millisievert per year averaged over five years (*i.e.* 100 millisievert in 5 years) with a maximum of 50 millisievert in any one year. For public exposure, 1 millisievert per year averaged over five years is the limit. In both categories, the figures are over and above background levels, and exclude medical exposure.

Post-accident Exposure

These low exposure levels are achievable for normal nuclear power and medical activities, but where an accident has resulted in radioactive contamination their application has no net health benefit. There is a big difference between what is desirable in the normal planned operation of any plant, and what is tolerable for dealing with the effects of an accident. Here, restrictive dose limits will limit flexibility in managing the situation and thus their application may increase other health risks, or even result in major adverse health effects, as near Fukushima since March 2011. The objective needs to be to minimize the risks and harm to the individual and population overall, rather than focusing on radiation in isolation.

This is recognised to some extent in the occupational health limits set for cleaning up such situations: The IAEA sets 100 mSv as the allowable short-term dose for emergency workers taking vital remedial actions, and 500 mSv as allowable short-term dose for emergency workers taking life-saving actions. At Fukushima, 250 mSv was set as the allowable short-term dose for workers controlling the disabled reactors during 2011. Following NRA consideration of the Fukushima experience, as well as overseas standards and the science, 250 mSv is now the proposed allowable dose in emergency situations in Japan from April 2016.

But even these levels are low, and there has been no corresponding allowance for neighbouring members of the public – ALARA was the only reference criterion regardless of its collateral effects due to prolonging the evacuation beyond a few days. The death toll and trauma from prolonged evacuation at Fukushima were clearly very much greater than the risks of elevated radiation exposure after the first few days.

This led to the IAEA in May 2013 publishing allowable dose rates for members of the public living normally in affected areas, measured 1m above the contaminated ground. A level of 220 mSv/yr over a full year is "safe for everyone" providing any ingested radioactivity is safe. Shorter term, at 40 times this level, 170 mSv for one week is provisionally safe, and at four times the yearly level – 880 mSv – is provisionally safe for one month.

In March 2011, soon after the Fukushima accident, the ICRP said that it "continues to recommend reference levels of 500 to 1000 mSv to avoid the occurrence of severe deterministic injuries for rescue workers involved in an emergency exposure situation." For members of the public in such situations, it recommends "reference levels for the highest planned residual dose in the band of 20 to 100 millisieverts (mSv)", dropping to 1 to 20 mSv/yr when the situation is under control.

After the Chernobyl accident people in Sweden were introduced to the concept of a becquerel budget relating to diet. An intake of 75,000 Bq of Cs-137 or 50,000 Bq of Cs-134 gives about 1 mSv of dose. Meat with up to 1500 Bq/kg was allowed to be sold. (Caesium does not bio-accumulate.)

When making decisions on evacuations, optimization of all the health risks (not just radiation exposure) is required. Rather than ALARA, a concept of As High As Relatively Safe – AHARS – would be more appropriate in dealing with emergency or existing high exposure situations, based on scientific evidence. This would be similar to the 680 mSv

Tolerance Dose allowable to 1950s (assuming gamma and beta radiation), and it would take into account the dose rate. This AHARS approach is supported by Allison and Cuttler among others, and an AHARS level of 1000 mSv/yr or 100 mSv per month is suggested. This would also mean that most or all of the displaced residents from near the Fukushima plant could return home, without any elevated cancer risk.

Another proposal is for As High as Naturally Existent – AHANE. That is, the highest naturally occurring radiation exposures in the world experienced by a large number of people without discernable negative health effects. At Ramsar in Iran, about 2000 people are exposed to at least 250 mSv/yr with no adverse effect. But in Guarapari, Brazil (pop. 73,000), Kerala, India (pop. 100,000), and Yangjiang, China (pop. 80,000) the average exposures are about 50 mSv/year, 38 mSv/year and 35 mSv/year respectively. In all cases, the residents have life expectancies at least as long as their national peers, and cancer rates slightly lower than fellow countrymen. US Navy shipyard worker data confirm the safety of 50 mSv/yr.

Nuclear Fuel Cycle Radiation Exposures

The average annual radiation dose to employees at uranium mines (in addition to natural background) is around 2 mSv (ranging up to 10 mSv). Natural background radiation is about 2 mSv. In most mines, keeping doses to such low levels is achieved with straightforward ventilation techniques coupled with rigorously enforced procedures for hygiene. In some Canadian mines, with very high-grade ore, sophisticated means are employed to limit exposure. Occupational doses in the US nuclear energy industry – conversion, enrichment, fuel fabrication and reactor operation – average less than 3 mSv/yr.

Reprocessing plants in Europe and Russia treat used fuel to recover useable uranium and plutonium and separate the highly radioactive wastes. These facilities employ massive shielding to screen gamma radiation in particular. Manual operations are carried by operators behind lead glass using remote handling equipment.

In mixed oxide (MOX) fuel fabrication, little shielding is required, but the whole process is enclosed with access via gloveboxes to eliminate the possibility of alpha contamination from the plutonium. Where people are likely to be working alongside the production line, a 25mm layer of perspex shields neutron radiation from the Pu-240. (In uranium oxide fuel fabrication, no shielding is required.)

Radiation Chemistry

Radiation chemistry involves the study of nuclear reactions. There are clear differences between a "chemical reaction," and a "nuclear reaction:" A chemical reaction involves electrons of an atom (which orbit the nucleus), while a nuclear reaction involves a reaction within the *nucleus* of an atom (which consists of protons and neutrons). Marie Curie discovered "radioactivity," which is the spontaneous disintegration of the unstable nuclei of some elements, or isotopes of elements, into more stable nuclei.

Recall that isotopes are different forms of an element which contain the same number of protons as the element on the periodic table, but each isotope of the element contains a different number of neutrons (thus isotopes have varying mass numbers). Generally, all isotopes with an atomic number over 83 are considered radioactive. Because a great amount of energy holds the nucleus together, in a nuclear reaction, a great amount of energy is released as the nucleus is disintegrated: this energy is called the "binding energy." The term "radiation" refers to the particles *and* energy that are released as a result of the radioactive reaction.

UNSTABLE NUCLEUS → MORE STABLE NUCLEUS + ENERGY

It is important to note that in a radioactive reaction, as in almost all reactions, the law of conservation of mass applies, which means that the total mass before the reaction has to equal the total mass after the reaction occurs. This concept is key in understanding radioactive reactions.

To understand the notation of nuclear reactions, it is helpful to review what "mass number," and "atomic number" refers to, and how they are represented along with the chemical symbol of an atom. Mass number refers to the total number of neutrons and protons in a nucleus, and atomic number refers to the total number of protons in the nucleus. Given a nucleus "X," mass number and atomic number will be denoted as follows:

$$^{\text{Mass number}}_{\text{Atomic number}}X$$

Also, an isotope of element "X" with mass number of 50, for example, may be written as "X-50."

There are three primary types of nuclear reactions: alpha, beta, and gamma radiation. These three main types of radiation are outlined below:

1. Alpha Decay:

In alpha decay, an unstable radioactive nucleus releases an "alpha particle," which is essentially a helium nucleus, and therefore the alpha particle has a mass number of 4 and an atomic number of 2, and can be denoted in either one of the two ways shown below:

$$^{4}_{2}He \text{ or } ^{4}_{2}\alpha$$

Because of the law of conservation of mass, the new nucleus has a mass number that is four less than that of the initial nucleus, and an atomic number that is decreased by two. In layman's terms, if you add the mass numbers on one side of the equation, they should equal the mass number on the other side of the equation, and the same goes for atomic numbers.

In an alpha radiation problem, often you are given only the initial unstable nucleus and asked to figure out the resulting nucleus. You know an alpha particle is released, so you use the law of conservation of mass to figure out the missing nucleus. Once you have found the missing nucleus, you can then use its atomic number to identify the element (referencing the periodic table):

$$^{227}_{88}Ra \rightarrow ? + ^{4}_{2}He$$

$$^{227}_{88}Ra \rightarrow ^{223}_{86}? + ^{4}_{2}He$$

Mass number of unknown nucleus: 227 – 4= 223

Atomic number of unknown nucleus: 88 – 2= 86

If we look on the periodic table, Radon (Rn) has an atomic number of 86.

Therefore, our complete reaction is:

$$^{227}_{88}Ra \rightarrow ^{223}_{86}Rn + ^{4}_{2}He$$

2. Beta Decay

In beta decay, an unstable radioactive nucleus releases a "beta particle," which is an "electron." An electron has no mass, thus it has no mass number, and essentially a -1 "atomic number," due to the -1 charge of an electron:

$$^{0}_{-1}\beta \text{ or } ^{0}_{-1}e$$

In much the same way as we solved the alpha decay reaction, we can solve for the new nucleus in beta decay,

$$^{216}_{84}Po \rightarrow ? + ^{0}_{-1}\beta$$

$$^{216}_{84}Po \rightarrow ^{216}_{85}? + ^{0}_{-1}\beta$$

- Mass number of unknown nucleus: 216 –0= 216.
- Atomic number of unknown nucleus: 84 – (-1)= 85.

If we look on the periodic table, Astatine (At) has an atomic number of 85.

Therefore, our complete reaction is:

$$^{216}_{84}Po \rightarrow ^{216}_{85}At + ^{0}_{-1}\beta$$

3. Gamma Decay:

In gamma decay, a high-energy photon is released. Radioisotopes, used in medicine for uptake in a person's body to detect key organs and glands, are frequently made by gamma reactions. A gamma particle is:

$$^{0}_{0}\gamma$$

Thus, in this type of radiation, there is no change in mass number or atomic number- the nucleus stays the same, and there is just a release of energy. Here is an example of a gamma reaction, again following the law of conservation of mass:

$$^{202}_{81}Tl \rightarrow ^{202}_{81}Tl + ^{0}_{0}\gamma$$

Other nuclear particles (besides the alpha, beta and gamma particles) seen in nuclear reactions include:

1) Neutron : $^{1}_{0}n$
2) Positron : $^{0}_{+1}e$
3) Proton : $^{1}_{1}H$

Uses of Radiation

One use of radiation is in dating fossils, or any specimen that you want to find the age of. The "half life" is the time required for half of the nuclei in a specific element to undergo radioactive decay. For instance, let's say:

The half life of I-123 is 20 hours. How much of a 100 mg sample of I-123 is left after 60 hours?

1. First, you want to figure out "how many 20 hrs are in 60 hours?" This would tell you how many half-lives the specimen goes through: 60 ÷ 20= 3 half lives.
2. So this means, that you have to start with 100 mg, and divide that in half three times, so: 100 mg, becomes 50 mg, 50 mg becomes 25 mg, 25 mg becomes 12.5 mg. So 12.5 mg of I-123 is left after 60 hours.

"Fission reactions" refer to the splitting of an atomic nucleus into approximately equal parts with a release of energy. Each subsequent nuclei that results can continue to be split in a type of "fission chain reaction," generating a great amount of energy. This is precisely how nuclear power plants generate a large amount of energy. About 15-20% of the world's energy comes from these nuclear power plants. "Fusion reactions" refer to reactions where multiple nuclei, combine to form a heavier nucleus (the opposite of fission reactions). The sun and stars undergo fusion reactions.

Also, irradiation of food by gamma rays from certain isotopes also increases shelf life of the food. For instance, irradiation of milk gives it a shelf life of three months without refrigeration.

Dangers of Radiation

Radiation can ionize atoms in our tissues, and cause reactions similar to decomposition or combustion in our skin. For instance, radiation can mutate nucleotide sequences in our DNA, potentially causing cancer.

The dangers of the specific type of radiation- alpha, beta or gamma- depend on each of their penetrative abilities. Alpha particles are not very dangerous because they cannot penetrate the barrier of the skin. Beta particles can penetrate about one centimeter deep in our skin. The most dangerous are gamma rays, which can penetrate through the skin very effectively.

Units called "REMs" are used to measure the effective biological damage that radiation causes. Some natural sources of radiation include cosmic radiation, the earth's radiation, and even the radiation from elements found naturally in human tissue. Medical sources of radiation include X-rays, and radiotherapy, for instance in cancer treatment. Artificial sources of radiation include the nuclear power industry, and fallout from nuclear tests. However, the average exposure for an individual with all of these sources combined is only about 0.2 REMs/year. Here is the extent of damage with greater exposure, due to living by a nuclear power plant, or a nuclear meltdown.

REMs	Damage
25	Changes in blood cell components.
200	In a short period of time, you will have "radiation sickness:" nausea, vomiting, decrease in white blood cell count, dehydration, diarrhea, hemorrhaging and loss of hair.
400 (anyone near the Chernobyl plant when it melted received this amount of damage almost immediately).	Half of the exposed population will be dead in sixty days.
600	All exposed to this level will be dead in a week.

Radiation Chemistry of Organic Solids

The effects of radiation on organic solids (polymers, lignocellulosic materials, etc.) are an area of increasing interest. Ionizing radiation can modify the physical, chemical, and biological properties of polymers; the changes occurring when polymer materials are in a solid state are different, as compared to chemical or thermal reactions carried out in melted polymers. After the interaction of ionizing radiation (gamma rays, X-rays, accelerated electrons, ion beams) with macromolecules, very reactive intermediates (primary entities, radicals and ions) are formed. These intermediates can follow several different reaction paths that result in the formation of oxidized products, grafts, scissioning of main chains (degradation) or crosslinking. The type and degree of these transformations depend on the structure of the material (e.g. polymer) and the treatment applied before, during and after irradiation.

Primary and Secondary Effects

Ionizing radiations can be classified in three categories:

- Charged particles: electrons (e^-), positrons (e^+), and heavy ions;
- Photons: gamma (γ), X-rays and light;
- Neutrons (n).

The term "ionizing radiation" in a wider sense also applies to photons or particles having sufficient energy to ionize a molecule of a material. This involves photons with energies ranging from the first ionization energy of the material (~10 eV) up to several million eV (MeV), as well as energetic charged particles, such as electrons, positrons, accelerated heavy ions, etc. The result of such energy absorption is the breaking or rearrangement of chemical bonds, i.e. decomposition of some of the initial material.

Radiation principles explain how the gamma rays, electron beams (EB) and X-rays interact with matter. Independently from the type of radiation (electromagnetic or particle) most of the radiation effects are produced by secondary electrons. According to calculations by Bethe with primary electrons of 10^{3-15} eV more than 80% of the absorbed energy is transferred to secondary electrons.

These interactions result in the formation of energetic electrons which are random throughout the materials and which cause the formation of energetic molecular ions. These ions may be subject to electron capture and dissociation, as well as to rapid rearrangement through ion-molecule reactions, or they may dissociate with time depending on the complexity of the molecular ion. The chemical changes in matter can occur via primary radiolysis effects, which occur as a result of the adsorption of the energy by the absorbing matter, or via secondary effects, which occur as a result of the high reactivity of the free radicals or electrons and the excited ions or molecules produced as a result of the primary effects. These highly reactive intermediates can undergo a variety of reactions leading to stable chemical products. In general, these chemical products can be detected and are referred to as radiolysis products. Understanding gamma-ray interaction with matter is important from the perspective of shielding against their effects on biological matter. These chemical changes can ultimately have biological consequences in the case where the target materials include living organisms.

The main types of interactions of gamma and X-radiation with substances are the photoeffect both in its photoelectric and photonuclear forms, Compton effect and scattering and generation and/or annihilation (elastic collisions) of the pairs (electron positron pair production). To a minor extent, photofission, Rayleigh scattering and Thompson scattering also occur. The photoelectric effect can displace atomic electrons, whereas the photonuclear reaction would displace elementary particles from the nucleus.

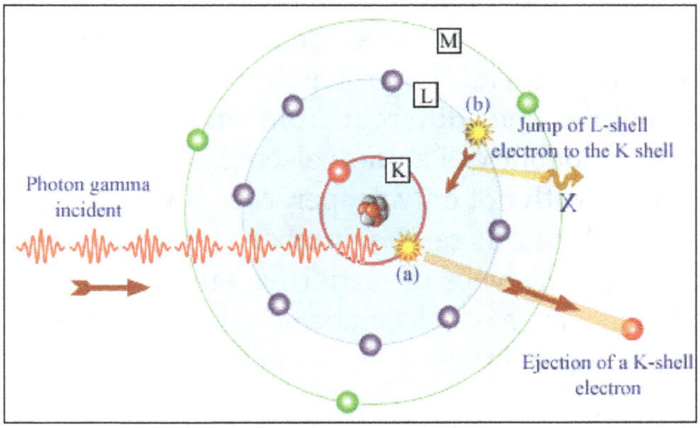

Gamma absorption by an atom: Schematic representation of photoelectric effect.

Photoelectric effect is the most effective mechanism of photon absorption. The photoelectric effect occurs in two stages. First, a photon (a) removes a bound electron in one atom. In the case of gamma photons, it is usually an electron belonging to the innermost atomic shells, L or K (as shown in Fig.). Then the atom that has lost one of its inner electrons is left in an excited state. An electron from an outer layer ((b) in Fig.) moves to occupy the vacancy left by the ejected electron. If the ejected electron belonged to the K shell, an X-ray is emitted during this transition which is accompanied by the emission of a characteristic soft electromagnetic radiation in the X-ray, ultraviolet (UV) or visible region of the electromagnetic spectrum.

The photoelectric effect is the phenomenon that transforms light, infrared and ultraviolet rays into electricity in solar panels and photodiodes of cameras. It is also involved in the completely different fi eld of radioprotection by transforming penetrating X- and gamma rays into electrons easy to stop and thus protects humans from the effects of these radiations.

The photoelectric process is always accompanied by a secondary emission since the atom cannot remain in an excited state indefinitely, thus:

- The atom emits X-rays and returns to the ground state.
- Auger electrons are emitted from the outer electronic shells carrying out the excitation energy. The secondary radiation is also later absorbed and occurs in scintillators used in gamma radiation detection.

The photoelectric effect is the most effective physical phenomenon in mitigating these radiations. The gamma-ray or X-ray photon absorbed by interacting with a bound electron to form an atom radical (•) disappears. The photon simply vanishes. If the energy of the photon is greater than 0.339 MeV, an electron may be ejected from the atom producing an ion pair. The ejected electron will then produce secondary ionization events in its surrounding atoms in a similar manner to beta particles (electrons or positrons). The photoelectric effect has a high probability of occurring with lower energy photons and atoms having high atomic numbers. The shell structure of atoms plays a crucial role. The photon removes an electron only if its energy exceeds the binding energy of the electron on its shell. The probability (called cross section) of removing an electron from a shell becomes different from zero beyond this threshold. Where there is a certain small probability of a photoelectric effect type phenomenon, gamma rays tend to interact more with not only deeper, core-level energy states, but also with the nucleus of the material. In this case, a number of different phenomena go on such as nuclear excitation, different types of scattering, etc. Photonuclear reactions can be used to produce neutron sources which can be used in a variety of applications such as nuclear medicine and radiography.

In the Compton effect, the photon undergoes a collision with an atom thus being deviated from its trajectory and releasing part of its energy. This energy is usually enough to abstract an electron from the atom. Compton scattering is the inelastic scattering of a photon by a charged particle, usually an electron. It results in a decrease in energy (increase in wavelength) of the photon (which may be an X-ray or gamma-ray photon). This is called the Compton effect. The electrons in matter are neither free nor at rest. However, in some cases, one can approximate the state of an electron in a simple model as free and at rest. In this case a gamma rays can interact with a loosely bound electron by being scattered with an appropriate loss in energy, as illustrated in figure.

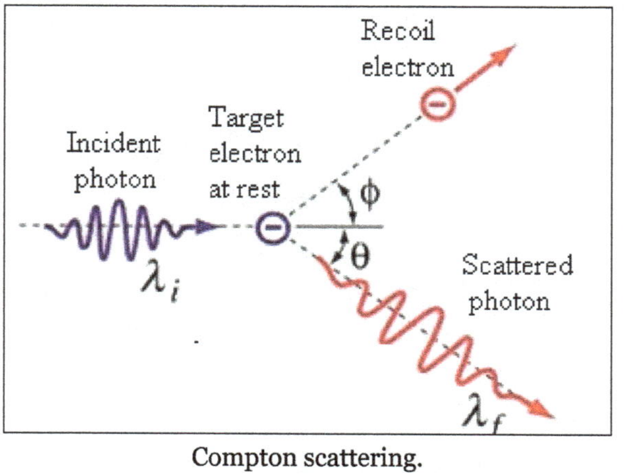

Compton scattering.

The scattering of photons from charged particles is called Compton scattering after Arthur Compton, who was the first to measure photon-electron scattering in 1922. When the incoming photon gives part of its energy to an electron, then the scattered photon has lower energy and according to the Planck relationship has a lower frequency and longer wavelength. The wavelength change in such scattering depends only upon the scattering angle for a given target particle. The constant in the Compton formula can be written as in above equation and is called the Compton wavelength for the electron:

$$\lambda_f - \lambda_i = \Delta\lambda = \frac{h}{m_0 c}(1-\cos\theta)$$

$$\frac{h}{m_e c} = \frac{hc}{m_e c^2} = \frac{1240\,\mathrm{eV\cdot nm}}{0.511\,\mathrm{MeV}} = 0.00243\,\mathrm{nm}\left(2.43\times10^{-12}\,\mathrm{m}\right)$$

Where:

λ_i – the initial wavelength,

λ_f – the wavelength after scattering,

h – the Planck constant,

m_o – the electron rest mass,

c – the speed of light, and

θ – the scattering angle.

Part of the energy of the photon is transferred to the recoiling electron. Inverse Compton scattering also exists, in which a charged particle transfers part of its energy to a photon. Compton scattering is of prime importance in radiobiology, as it is the most probable interaction of gamma rays and high energy X-rays with atoms in living cells and is applied in radiation therapy. In material physics, Compton scattering can be used to probe the wave function of the electrons in matter. The primary effect of the Compton interaction is ionization.

Chemical Effects of Radiations

The primary effect of high energy radiation is ionization and, to a minor extent, excitation of molecules and atoms. The radiation gradually releases its energy through the material giving rise to a track of reactive species (ions, excited molecules, radicals, electrons) whose reactions are responsible for the irreversible chemical transformations of the matter (radiolysis). Primary ionization generates electrons, whose energy is large enough to give rise to further ionizations and excitations. Thus, secondary electrons are responsible for a major part of radiation chemical effects. This explains the substantial analogy of the effects stemming from the different type of radiations either electromagnetic (gamma rays, X-rays) or charged particles (electrons, alpha, protons, deuterons, etc.). The probability that electrons with energy < 10 MeV and gamma radiations of 0.5-1 MeV (Compton effect) will interact with atoms and molecules is proportional to the electron density of the material and it is therefore relatively unselective and largely independent of chemical structures. A general radiolysis mechanism for a condensed phase is illustrated in figure.

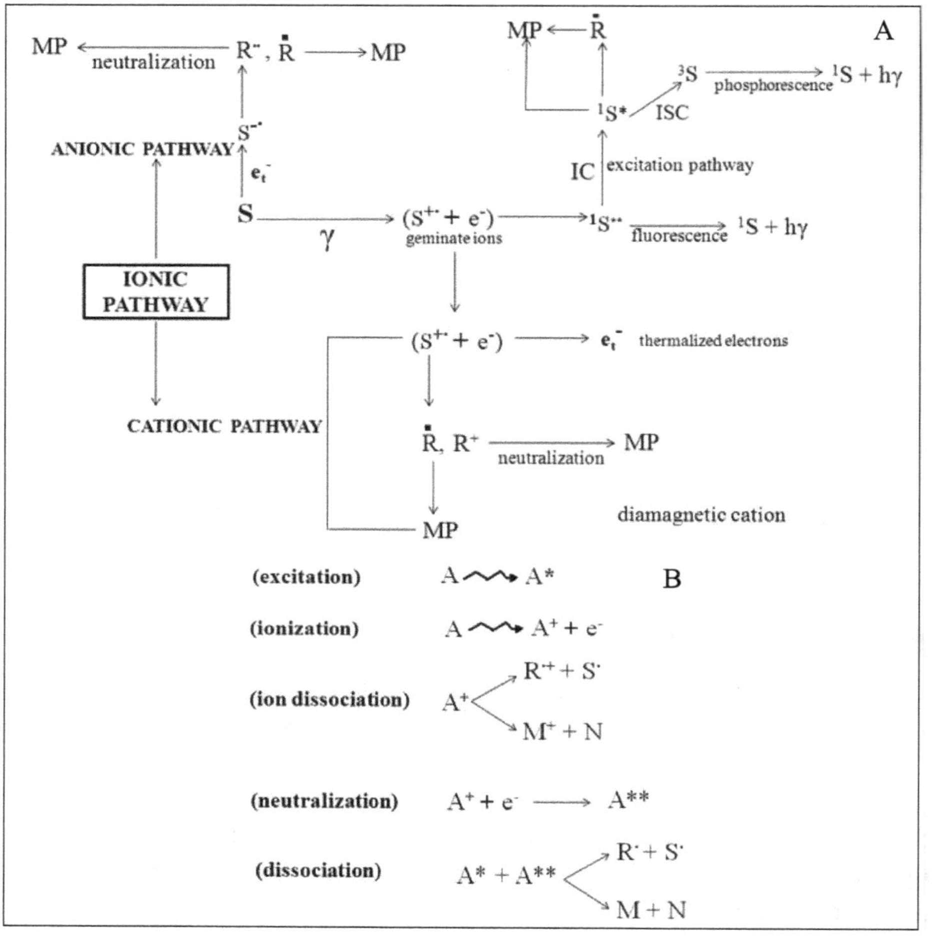

In figure, General radiolysis scheme (A) and primary processes in the radiolysis of organic compounds (B): $R^{\cdot}$ and $S^{\cdot}$ – free radicals; $S^{\cdot+}$ and $S^{\cdot-}$ – cation or anion radicals, respectively; ISC – intersystem crossing; IC – internal conversion; R^{+} and R^{-} – diamagnetic cation or anion, respectively; MP, M and N – molecular products.

This consists of excitation radiolysis and ionic radiolysis pathways depending on the type of active species generated. The free radicals are the most important intermediates in all radiolysis processes. The excited species in condensed systems can undergo: (i) Fast internal conversion (10^{-13} s) to the lowest excited levels with degradation to thermal energy of most of the electronic excitation energy, (ii) Chemical reactions, (iii) Return to the fundamental state of the same multiplicity with emission of radiation, (iv) Fluorescence or (v) Return to the fundamental state after intersystem crossing (ISC) with emission of phosphorescence radiation. The primary processes in the radiolysis of or ganic compounds can, in most cases, be represented by excitation, ionization, ion dissociation, neutralization and dissociation.

Macromolecules are very sensitive to minute chemical changes at the appropiate places which have great influence on the morphology. A single bond altered may convert a free flowing polymer from a fluid to an elastic solid. An even smaller change may be lethal in biological systems.

The primary interactions of ionizing radiation with polymers include ionization, stabilization of electrons through generation of hot electrons, ion neutralization and free radicals formation. Free radicals are created either through chain scission or through the dissociation of the side chain (C–H(Cl)) bonds.

Table: Primary processes occurring in polymer irradiation (P – polymer, R• – radical).

Ionization:	$P \rightsquigarrow P^{+} + e^{-}$
Excitation:	$P \rightsquigarrow P^{*}$
Thermal stabilization of electron:	$e \rightarrow e^{-}$
Neutralization:	$P^{+} + e^{-} \rightarrow P^{*}$
Free radicals formation:	$P^{*} \rightarrow R_1^{\bullet} + R_2^{\bullet}$
Disproportionation:	$\rightarrow R^{\bullet} + H^{\bullet}$

The irradiation of polymers, natural or synthetic, with ionizing radiation leads to the formation of reactive intermediates, free radicals, ions and atoms and molecules in excited states. These intermediates can undergo several reactions such as hydrogen abstraction, rearrangements and formation of new bonds, addition to the double bonds, chain scission, oxidation and grafting recombination (crosslinking and branching, to an extent which depends upon the polymer structure).

Table: Secondary reactions in polymers occurring under irradiation (M – monomer).

Hydrogen abstraction:	$R^{\bullet} + P \rightarrow RH + P^{\bullet}$
Addition to the double bonds:	$RCH{=}CH_2 + H^{\bullet} \rightarrow RCH_2\,CH_2^{\bullet}$
Recombination	
(branching or/and crosslinking):	$R_m^{\bullet} + R_n^{\bullet} \rightarrow P_{m+n}$ or/ and 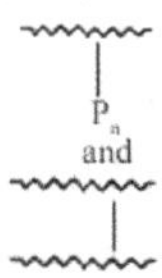
Chain scission:	$R_m^{\bullet} \rightarrow R_{m+n}^{\bullet} + R_n^{\bullet}$
Oxidation:	$R^{\bullet} + O_2 \rightarrow ROO^{\bullet}$
	$ROO^{\bullet} \rightarrow ROOH, -C{=}O, -OH, -COOH$
Grafting:	$R^{\bullet} + M \rightarrow RM^{\bullet}$
	$R^{\bullet} + nM \rightarrow RM_n^{\bullet}$

As a result both low molecular weight species and crosslinking occur with changes in molecular weight (decrease or increase, correspondingly). Both reactions can occur simultaneously, one of them being dominant depending on polymer structure, the presence of additives, the radiation dose and environmental conditions.

Crosslinking dominates under vacuum, while chain scission may be a main reaction in presence of oxygen or air. As a consequence, radiolytic products from both a polymer and its additives are formed. As a rule, under the radiation effect on solid substances, the spatial distribution of radicals is non-uniform. Sometimes it may present closely located radical pairs and/or agglomerations. Trapped radicals increase with irradiation dose.

The fundamental processes related to these reactions include:

- Crosslinking – Polymer chains are joined and a network is formed;
- Chain scission – The molecular weight of the polymer is reduced through chain scission;
- Oxidation – Where the polymer molecules react with oxygen via peroxide radicals (oxidation and chain scission often occur simultaneously);
- Long-chain branching – Polymer chains are joined but a three-dimensional network is not yet formed;
- Grafting – A new monomer is added onto the base polymer chain;
- Polymerization – Can also be initiated when monomers are irradiated;
- Radiation curing (as in the case of coatings or composites) is a combination of polymerization and crosslinking.
- Different polymers have different responses to radiation, especially crosslinking vs. chain scissioning. The same polymer such as poly(tetrafl uoroethylene) can degrade by chain scission when $T < T_m$ (melting temperature of 315°C) or it crosslinks when $T > T_m$. Frequently different reactions occur simultaneously with predominance of some of them.

Radiation Yield

To quantify the radiation chemical effects of ionizing radiation, the number of molecules transformed or produced when a certain the quantity of radiation energy is absorbed should be known. The G(X) value is a measure of the radiation-chemical yield. It was originally defined as the number of molecules n(X), produced, destroyed or changed by radiation (consumed) to the irradiated matter by the mean energy imparted, E, usually of 100 eV of absorbed energy.

$$G(X) = \frac{n(X)}{E}$$

Hence, the unit of the G-value of the International System of Units (SI) is mol/J. In the earlier literature the G-values were always given in units molecules/100 eV, but the new SI units will be used. (1 molecule/100 eV corresponds to 1.036×10^{-7} mol/J).

Table: G-values for crosslinking and chain scission for some common polymers.

Polymer	Crosslinking G(X)	Scission G(S)	G(S):G(X)
Low-density polyethylene	1.42	0.48	0.34
High-density polyethylene	0.96	0.19	0.20
Isotactic polypropylene	0.16-0.26	0.29-0.31	1.1-1.5
Atactic polypropylene	0.4-0.5	0.3-0.6	0.7-0.9
Poly(methyl methacrylate)	< 0.50	1.1-1.7	> 2
Poly(tetrafl uoroethylene)	0.1-0.3	3.0-5.0	10
Natural rubber	1.3-3.5	0.1-0.2	0.14
Nylon-6	0.35-0.7	0.7	1.0
Nylon-6,6	0.5-0.9	0.7-2.4	1.4
Poly(vinyl acetate)	0.1-0.3	0.06	0.2
Poly(vinylidene fl uoride)	0.6-1.00	0.30-0.6	0.3
Poly(methyl acrylate)	0.45-0.52	0.08	0.15
Polystyrene	0.019-0.051	0.0094-0.019	0.4
Polybutadiene	5.3	0.53	0.10
Polyisobutylene	0.05-0.5	5	> 10
Butyl rubber	< 0.5	2.9-3.7	> 6

Table lists the G-values for crosslinking, G(X), and chain scissioning, G(S), for some of the common polymeric materials irradiated at room temperature without oxygen. Materials with G(X) larger than G(S), the overall result is crosslinking, and whenever G(S) is larger than G(X), the overall result is degradation. Materials whose G(X) and G(S) values are both low are more resistant toward radiation.

Radiation effects are remarkably sensitive to molecular structure. G(S) and G(X) values vary from 0.1 to 10, as shown in table. The ratio between them determines the changes in average molecular weight of polymers.

The G(X) of polyethylene (PE) is about one order of magnitude larger than that of polystyrene (PS) which demonstrates that aromatic rings exert a protective radiation effect.

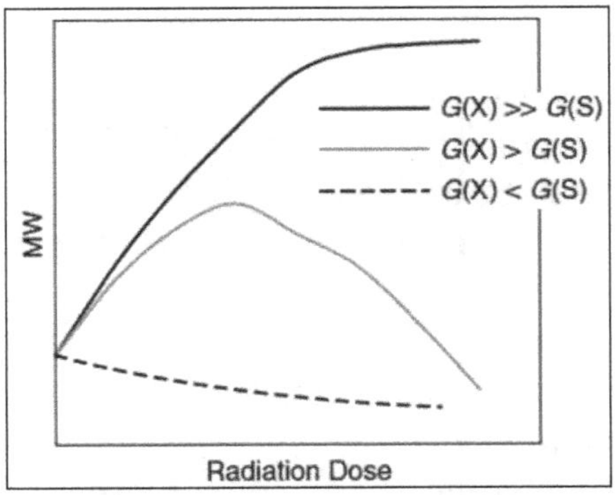

Relation between average molecular weight of polymers, G(X) and G(S) and radiation dose.

G(X) and G(S) also depend on irradiation conditions, such as temperature and atmosphere.

Radiolysis Products

Polymers used in medical devices and in primary packaging materials have been studied with respect to radiolysis products. Orthopedic implants and other medical devices that come into close contact with the human body are sterilized by either gas or high energy radiation. They are not free from the effect of ionizing radiation particularly if the process is performed in air. Food irradiation, sterilization of packaging materials for pharmaceutical or food products with ionizing irradiation can cause not only chemical and physical changes in polymer materials but also formation of some radiolysis products which can be solvent extractable or detectable by other techniques such as gas chromatography/ mass spectroscopy (GC/MS), electron paramagnetic resonance (EPR), spectroscopic methods, in surrounding media. The products and their concentrations are characteristic for each polymeric material. The polyolefin materials (polyethylene, PE, and polypropylene, PP) show an increase in low volatile compounds after irradiation due to an oxidative decomposition of the polymer and of additives. PE produces only traces of hydrocarbons, aldehydes, ketones, and carboxylic acids, which largely disappear within weeks. Other packaging materials, such as poly(ethylene terephthalate) (PET), polyamide (PA), and polystyrene (PS), did not significantly change their amount of solvent extractable compounds after irradiation with 25 to 44 kGy. PA-6 (nylon-6) yields pentanamide as the main product, plus traces of some homologous amides.

The poly(vinyl chloride) (PVC) packaging material release of hydrochloric acid (HCl) during irradiation and large amounts of volatile substances were extracted from the PVC sheets. The main products of PVC and PP are fragments of additives, i.e. of stabilizers and phenol-type antioxidants, respectively. For consumer protection and also to meet general food packaging legislative requirements for irradiated packaging materials, it is necessary to evaluate the compositional changes in the polymers following irradiation, especially for irradiated polyolefins and PVC. Most attention should be paid to low volatile radiolysis products which are the most likely to migrate into a foodstuff or a pharmaceutical products. Under realistic polymer/food simulant contact conditions during irradiation, a large number of primary and secondary radiolysis products (hydrocarbons, aldehydes, ketones, alcohols, carboxylic acids) were produced. These compounds were detected in the food simulants after contact with all fi lms tested, even at the lower absorbed doses of 5 and 10 kGy (approved doses for food preservation). The type and concentration of radiolysis products increases progressively with increasing dose. Volatile and non-volatile compounds produced during irradiation affected the sensory properties of potable water after contact with packaging films. Taste transfer to water was observed mainly at higher doses and was more noticeable for multilayer structures containing recycled low density polyethylene (LDPE). The main products of PS are acetophenone, benzaldehyde, phenol, 1-phenylethanol, and phenylacetaldehyde. Their concentrations are one order of magnitude below the residual styrene/styrene dimer levels.

The radiolysis products formed during gamma-ray or electron beam radiation are generally later released. Due to rigid structure of these materials, the radiolysis products may be trapped inside the polymeric structure and be delivered later when the material enters into contact with environment that favours their release. The radiolysis products of some polyethylene irradiated at 20 kGy were evaluated 7.5 months after irradiation and it was found that the peroxy radicals continued to react, promoting chain scissioning, and the generation of carbonyl, of hydroperoxides and of ketones.

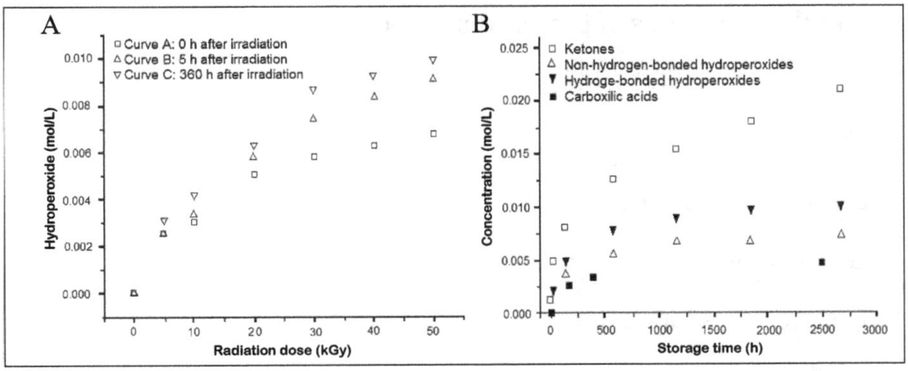

Variation of functional groups with radiation dose (A) and storage time of UHMWPE (B).

In the radiation of ultra-high molecular weight polyethylene (UHMWPE), high concentrations and new radiolysis products were found.

When exposed to 25 or 50 kGy radiation doses, PS produces radiolysis products such as benzaldehyde, acetophenone (30-50 ppm), 1-phenylethanol (< 10 ppm) and phenol. Nylons evolved pentanamide (75 ppm) and caprolactam. PP generates alkyl radicals which can subsequently interact with molecular oxygen to form hydroxyl, carbonyl or carboxyl groups. These volatiles can have adverse flavour effects to irradiated food packaged in these polymers, in general in this order: LDPE > HDPE (high density polyethylene) > PS ~ PA and PET. The intensity of specific off-odors increases with the availability of oxygen in atmosphere.

The addition of the antioxidants reduces the formation of carboxylic acid derivatives in LDPE films. However, phenolic stabilizers, such as Irganox 1076, 1010 and 1330 and the arylphosphite antioxidant Irgafos 168 (in PVC, PE and PP) are susceptible to degradation under gamma irradiation. Low doses of irradiation (5 kGy) to HDPE trays can completely destruct phosphite antioxidants to their phosphate products with some covalently bonded on the polymer matrix so that they will not be found in simulants. At present, little is known about toxicity of radiolytic products from polymers and additives because of the variability of migration rates and because of the potential of polymeric entrapment of volatiles from the different polymers into food simulants.

Changes in Physical and Chemical Properties

Synthetic Polymers

Radiation is the unique source of energy which can initiate chemical reactions at any temperature, including ambient, under any pressure, in any phase (gas, liquid or solid), without use of catalysts. Polymers are quite often irradiated for modification or sterilization (medical products). Therefore, the changes in their structure may be beneficial or undesirable. The application of radiation for modification of synthetic materials, mostly curing and crosslinking is a well-established technology.

The changes in the properties of the irradiated polymers in the solid state are:

- Chemical changes:
    - Crosslinking and chain scission,
    - Structural weakness,
    - Gas formation,
    - Double bond formation,
    - Oxidation,
    - Living radical effects after irradiation,
    - Effects on additives.
- Physical changes:
    - Changes in mechanical properties (mechanical strength, flexibility, Young's moduli, torsional resistance, etc.),
    - Colour change,
    - Changes in conductivity,
    - Changes in crystallinity,
    - Thermal stability and transitions.

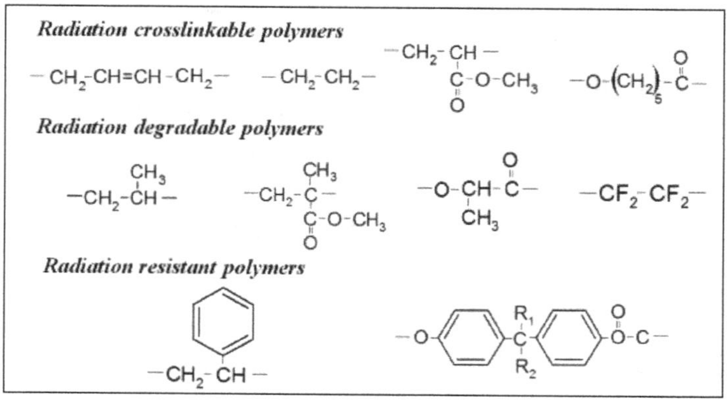

Examples of chemical structures of polymers with different responses to radiation.

The different responses to radiation for different polymers are related to the chemical structures of the polymers. Some examples of chemical structures that correspond to crosslinking types, degrading or scissioning types, and radiation-resistant polymers are given.

The formation of reactive oxygen species can result in the oxidation of polymer with the formation of peroxide, alcohol and carbonyl compounds. The most stable polymers are vinyl derivatives, PS and the PET with polyamides (nylons) having intermediary stability while polyolefins as the least stable polymers.

In solid state, the radicals can be found for months after irradiation. Commercial polymers often contain some additives. The effect of the additives on radiolysis is important. Radiation protecting additives have aromatic cyclic structures and reduce degradation by radiation. Plasticizers can have a similar protective role.

Mechanical properties of irradiated polymers depend on the ratio of chain scission and crosslinking and are related to molecular weight. The mechanical properties of polymers of the crosslinking type are improved upon irradiation with low absorbed doses (up to 0.1 MGy). The elastic modulus, tensile strength and hardness are increased, whereas the solubility is decreased with increasing absorbed dose. At very high absorbed doses, these polymers become hard and brittle. The polymers of the degradation type undergo deterioration of their mechanical properties even at low absorbed doses. Low doses of 5 and 10 kGy had no effect on mechanical properties such as tensile strength, elongation at break and Young's modulus. However, at 30 kGy, there was a decrease in tensile strength in HDPE, PP and a decrease in elongation at break of LDPE and a polyethylene ionomer. The mechanical properties of neither an ethylene-vinyl acetate (EVA) nor polystyrene were affected by irradiation.

Important aromatic polymers such as polyimides and polyarylsulphones, as shown in figure are resistant to high energy radiation compared to nonaromatic polymers.

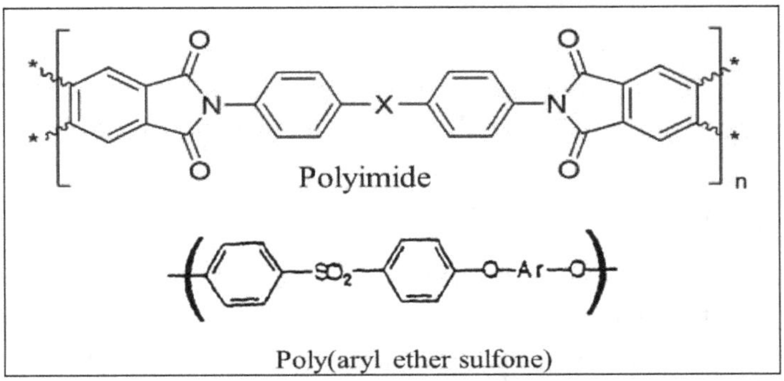

General chemical structure of polyimide and polysulphone.

The mechanical and electrical properties of a poly (pyromellitimide) remain satisfactory in the presence of air up to absorbed dose of 100MGy. Most aromatic polymers proved to be radiation resistant only in the absence of oxygen. For example, the mechanical properties of polystyrene were not appreciable deteriorated when exposed in the absence of oxygen to the absorbed doses up to 10 MGy, but these properties decreased to 50% of their initial values at an absorbed dose of only 0.8 MGy in the presence of oxygen. Irradiation of prepackaged meat and poultry should alter neither physicochemical properties of a packaging film nor result in the transfer of components or residues from packaging material to contaminate the food in contact with the plastic film. The irradiation of plastic film results in a combination of chemical crosslinking and a consequent increase of tensile strength of the film, or a fragmentation leading to a decreased strength and increased permeability of packaging films. In addition to these events, the formation of volatile radiolysis products (hydrogen, methane, HCl, etc.) is influenced by presence of oxygen in the irradiated products. It was reported that irradiation at 5, 10 and 30 kGy had no effect on the O_2, CO_2 permeability or water vapour transmission rates of a variety of polymers including LDPE, HDPE, PS, EVA, biaxially oriented PP and an ionomer.

Colour change can appear when the double bonds formed on irradiated polymers such as poly (methyl methacrylate) (PMMA) and PVC.

Conductivity change can increase when the double bond increased in irradiated polymers.

The overall migration into food simulants as distilled water was not affected by low irradiation doses; at 30 kGy, migration into 3% acetic acid was decreased for PP and an ionomer.

Natural Polymers

Studies of the radiolysis of biopolymers serve the dual purposes of giving information on: (1) the chemical mechanisms by which radiation modifies life processes and (2) the structure-properties-application relationships in macromolecules. Irradiation effects on biopolymers result in the cleavage of chemical bonds. Because of the very complex composition of most biopolymers, many different chemical reactions are initiated and in each case a host of products are formed under the influence of high energy radiation. A variety of biopolymers composed of linear strands, including certain polysaccharides, proteins and nucleic acids undergo predominantly main-chain scission, and this is of outstanding importance in the case of deoxyribonucleic acids (DNA), because the killing of living cells correlates well with double-strand breaks.

Amino Acids, Proteins and DNA

Amino acids and proteins exist as zwitterions. Under irradiation amino acids undergo by the electron capture deamination, as indicated in figure. With proteins both deamination and decarboxylation take place. For those molecules containing sulphur groups, they are cleaved by a radiolytic mechanism and hydrogen sulphide is produced as a by-product, as shown in figure.

In living cells, both direct and indirect depositions of energy are possible. Direct effects in chromatin components result in the formation of specific radical products, many of which are highly reactive. Secondary reactions of the cationic radicals are largely unknown. Indirect effects occur when energy is deposited in water or other components in a solution, and radiolysis products such as e^-_{aq} and •OH react with the biopolymer have been investigated, as shown in figure.

Commonly DNA molecules are tightly associated with 8 to 12 water molecules per nucleotide forming a primary solvation shell which increases the radiation-induced damage to DNA by ~50%. When the water molecules are ionized, an entire transfer to DNA can occur and electrons ejected from water molecules can be scavenged by DNA molecules. The events due to the direct effect are single- and double-strand breaks, crosslinking, base release, and lesions in bases and sugar moieties. Hydrogen bonds between the two strands of the double helix are broken. A large number of radiolysis products results. From base damage the following products have been identified: 5-hydroxy-5,6-dihydrothymine, 5,6-dihydrothymine, 5-hydroxyuracil, 5,6-dihydroxyuracil, 5-hydroxycytosine, 2-hydroxyadenine, 8-hydroxyadenine, 7-hydro-8-oxo-guanine, 2,6-diamino-4-oxo-5-formamidopyrimidine, 4,6-diamino-5-formamidopyrimidine.

Studies carried out half a century ago showed that this process is effective in inactivating enzymes. Replacing •OH to the less reactive inorganic radical anion $•Br_2^-$ has been done. It is then possible to determine the role of tyrosine in functional and structural integrity of several proteinase inhibitors.

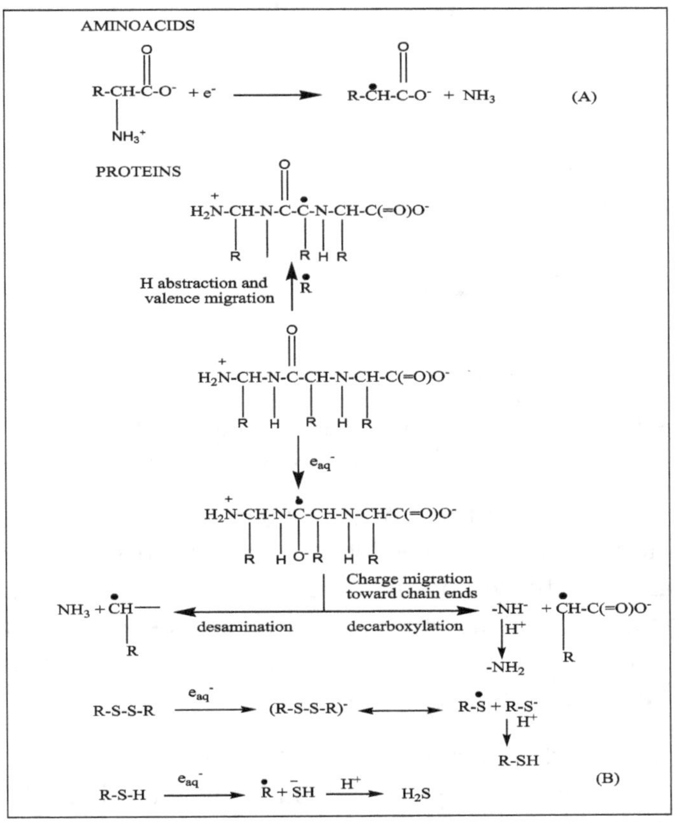

Radiolytic mechanism of amino acids (A) and proteins (B).

Radiolytic mechanism of DNA:

$$\text{Direct effect: } DNA \xrightarrow{\text{radiation}} DNA^+ + e^- + DNA^*$$

$$\text{Indirect effect: } H_2O \xrightarrow{\text{radiation}} H_2O^+ + e^-$$

$$H_2O + DNA \rightarrow H_2O + DNA^+$$

$$e^- + DNA \rightarrow DNA^-$$

Both e^-_{aq} and –OR react rapidly with DNA, but only •OH initiates reactions which damage DNA. Radiolysis of double-stranded DNA leads to an increase in optical absorption. The •OH is believed to attack the deoxyribose moiety, causing strand breaks and partial denaturation, thus reducing the hypochromic effect.

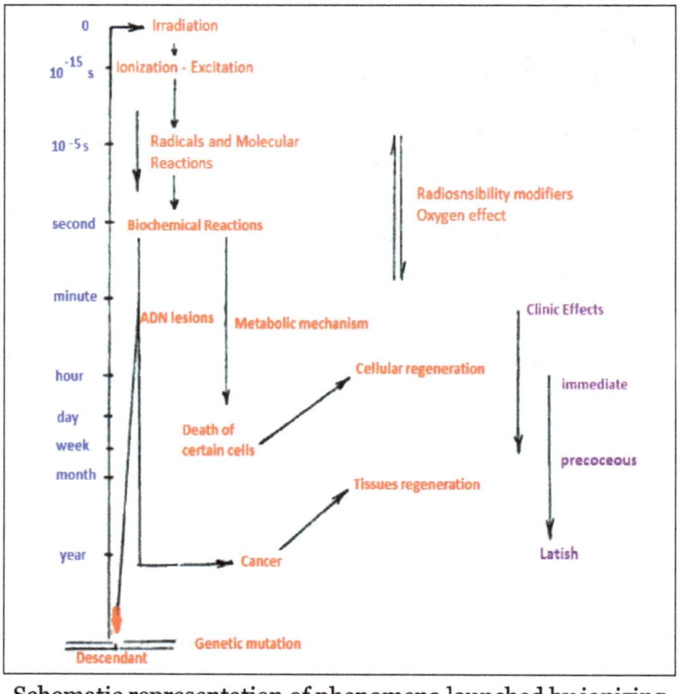

Schematic representation of phenomena launched by ionizing radiations in biological molecules/organisms.

After the DNA is partially denatured, or single-stranded, •OH attacks the bases also. Three kinds of strand breaks have been observed: (i) immediate, (ii) those appearing post irradiation, and (iii) those appearing on post-irradiation treatment with alkali. Radiolysis of chromatin results in DNA strand breaks, base damage, and protein-DNA crosslinks. Yields for strand breaks and base damage are lower in chromatin than in purified DNA, and lower still in intact cells.

In living organisms, this denaturation of the biological molecules without adequate protection over a long period of exposure could have deleterious effects on health, as illustrated in figure.

Polysaccharides

Irradiation of polysaccharides in the solid state induces the radical formation in molecular chains as a result of the direct action of radiation. Here mainly two events take place: (1) direct energy transfers to the macromolecule to produce macroradicals and (2) the generation of primary radicals due to the presence of water (moisture). The course of the degradation of carbohydrates in the solid state is illustrated in figure. The main effects are fragmentation, hydrolysis and rearrangement leading to low molecular weight products. When polysaccharides such as starch, cellulose, chitosan, amylose and dextran are subjected to high energy radiation, a vast number of products is formed, such as gaseous H_2, CH_4, CO, CO_2 and low molecular mass substances such as water, formaldehyde, methanol, aldehydes, ketones and sugars.

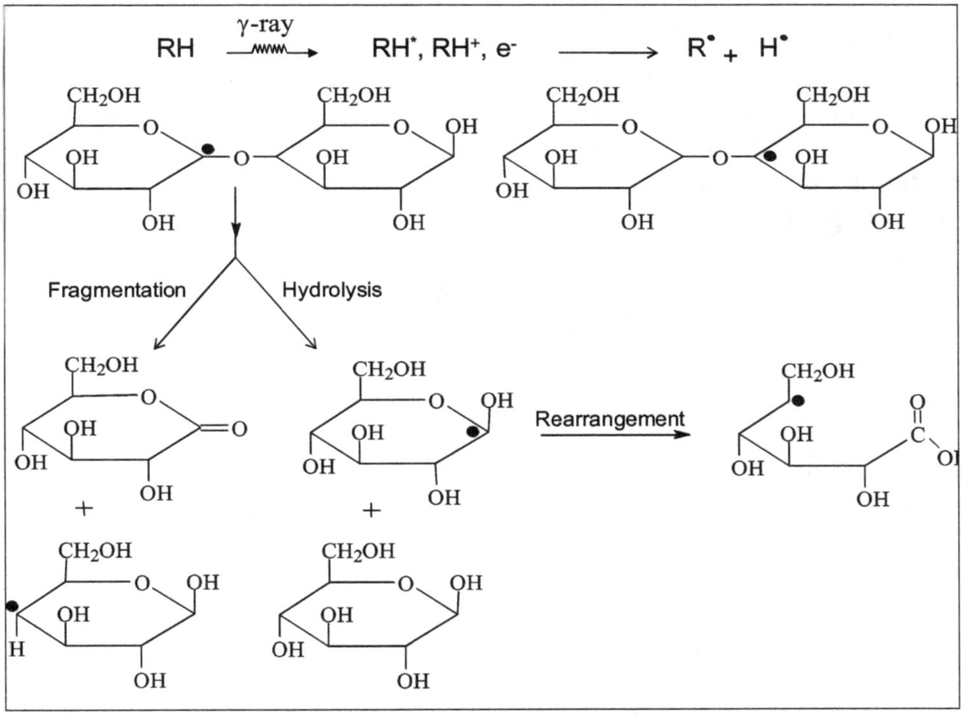

Events in solid state radiation of carbohydrates.

The remaining polymer contains carbonyl, carboxyl, and aldehyde groups. The average molecular molar mass is reduced because of the predominant main-chain scissioning through cleavage of the C–O–C glycoside bond between base units. When high energy radiation is directly absorbed by polysaccharides, electronically excited moieties, radical cations and electron are generated. Free radical processes play an important role, while the radical cations generated are rapidly transformed to free radicals by deprotonation.

During the solid state radiolysis of polysaccharides, scissioning of the glycosidic bond, analogous to the reactions shown in figure is the dominant process which eventually leads to a decrease in the molecular weight of the macromolecules. Degradation and the resulting changes in average molecular weight products are often quantified as radiation yields of degradation G(s). Moisture content strongly influences the reaction pathways and yields. The radiolysis of water contributes to this. The yields of radicals are significantly higher than in dry polysaccharide so the contribution of indirect effect may be greater than based on percent of moisture, and also due to the effect of water molecules on the dry matrix structure and polymer chain mobility.

Complex Materials

Complex materials, including multilayer laminated and coextruded ones, are very useful because they combine a number of desirable properties which no single of material possesses. Most are based on polyolefin, PS, PET, PA, EVA, and other synthetic or natural polymers. The effect of gamma irradiation on PP biocomposites (Eucalyptus globulus, pine cones, and Brassica rapa) has been studied by Parparita et al. The biomass lignin content acts as an antioxidant and has a stabilization effect against gamma irradiation degradation. For the materials studied, gamma irradiation promoted the biodegradability under the action of the Bjerkandera adusta fungus. Removing a protective layer of relatively inert polyolefin, increased the effects of absorbed dose which leads to increase of oxygen containing functional groups, as illustrated in figure. In this manner, an eco-friendly method for improving integration of some “undegradable” materials in natural cycles could be improved.

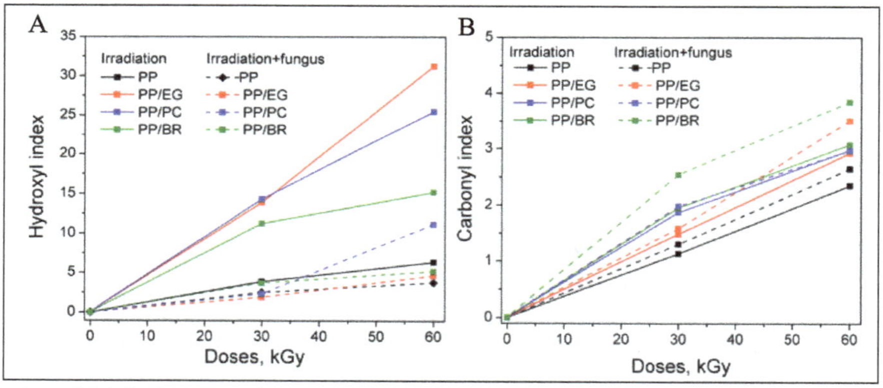

Variation of (A) hydroxyl and (B) carbonyl indexes of gamma-irradiated PP and PP/biomass composites compared with those of 7 weeks fungal degradation of the PP/biomass samples.

Radiation Protection

Measures to fully protect polymers against radiation exposure do not exist. The stabilization of polymeric materials against high energy radiation may be achieved through the use of antioxidants, ultraviolet and heat stabilizers, which, in turn, also undergo degradation. Radiolysis products could originate more from additive decomposition than from the originating polymers themselves.

Protective agents added to the polymers in small amounts are effective only to a limited extent and function via different mechanisms. They can act mainly as antioxidants by preventing radiation-induced oxidative chain reactions or more generally as free radicals scavengers. Other additives may act as energy absorbers in accepting electronic excitation energy transferred from a polymer. In this manner, agents that operate protectively are aromatic compounds, including polymers with aromatic constituents or fillers (inorganic materials or carbon black). Electronic excitation energy transported to or directly absorbed by aromatic rings is delocalized due to the π-electron-based resonant structure of the aromatic ring, and thus converted to heat by collision with neighbouring groups or molecules. Hence, bond rupture is much less likely in the case on nonaromatic compounds.

3

Radiochemistry and Radioactive Decay

The domain of nuclear chemistry which studies the chemical reactions and properties of non-radioactive isotopes using the radioactive isotopes of elements is called radiochemistry. Radioactive decay refers to the process through which an unstable atomic nucleus loses energy by emitting radiation. This chapter discusses in detail the concepts and theories related to radiochemistry and radioactive decay such as radioactive isotope and resonance fluorescence.

Radioactive Materials

Radioactive material includes substances or articles that are radioactive, or have become radioactive through non-natural processes, for example at a nuclear reactor.

There are two main types of radioactive substances:

- Open radioactive sources, also called unsealed sources.
- Sealed radioactive sources, also called closed sources.

Radioactive materials are used by many organizations and businesses. For example, hospitals, research organizations, radiographers and process industries use radioactive materials for:

- Diagnosing and treating disease.
- Controlling industrial processes.
- Preventing static electricity.
- Industrial radiography.
- Scientific research.

Different types of radioactive material are used for different activities.

Open Radioactive Sources

Open sources are radioactive materials that you can easily divide, disperse or dilute. They can be in a liquid, gaseous or sometimes solid form. Open sources include radioactive laboratory chemicals and radiopharmaceuticals. Open radioactive sources can potentially contaminate other material through leakage or leaching. You must store all open radioactive sources securely.

Sealed Radioactive Sources

Sealed sources have a structure which prevents radioactive material from leaking during normal use. If you use sealed sources incorrectly you could cause radioactive contamination.

Sealed radioactive sources are categorised according to their hazard or risk as:

- High activity sealed sources (HASS).
- Sources of similar potential hazard to HASS.
- Category 5 (low hazard or risk).

A sealed source may take the form of a welded steel capsule, or a homogenous, laminated, electrodeposited or foil source.

HASS present a greater hazard to the environment and human health than other sealed sources. HASS may include sterilisation sources, hospital radiotherapy sources, industrial radiography sources, density and moisture gauges and some industrial process control gauges.

Mobile Radioactive Apparatus

Mobile radioactive apparatus includes equipment, appliances or other things that are portable and classed as radioactive material. They may be used for:

- Testing, measuring or investigation.
- Releasing radioactive material into the environment or introducing it into organisms.

Mobile radioactive apparatus may contain sealed or open radioactive sources.

Radioactive Waste

Radioactive waste is anything that your business discards, intends to discard or is required to discard which contains radioactive material.

Radioactive waste can be in a solid, liquid or gaseous form.

There are four types of radioactive waste:

1. High-level radioactive waste that contains levels of radioactivity so high that heat is generated, e.g. waste created by reprocessing spent nuclear fuel.

2. Intermediate-level radioactive waste that contains more radioactivity than low-level waste but without the heat generation that occurs in high-level waste, e.g. redundant sealed sources.

3. Low-level radioactive waste that is within specified concentrations of radioactivity, e.g. everyday items such as paper or plastics contaminated by radioactivity.

4. Very low-level radioactive waste that is below a specific concentration of radioactivity, e.g. laboratory and medical equipment used to handle radioactive material.

Radioactive waste may include items contaminated by radioactive material or other radioactive waste, such as:

1. Clothing that needs to be disposed of;
2. Laboratory wastes from the use of open radioactive material;
3. A sealed source which is scrap.

Radioactive substances users at non-nuclear sites are only likely to produce very low-level or low-level radioactive waste.

Radiochemistry

Radiochemistry the branch of chemistry concerned with the chemistry of radioisotopes, elements, and substances, the laws governing the physicochemical behavior of this radioactive matter, the chemistry of nuclear transformations, and the physicochemical processes that accompany these transformations. Radiochemistry, because of the topics, methods, and objects of its investigations, can be subdivided into general radio-chemistry, the chemistry of nuclear transformations, the chemistry of radioactive elements, and applied radiochemistry.

General radiochemistry studies the physicochemical regularities in the behavior of radioisotopes and elements. Radioisotopes, which differ little in their chemical properties from nonradioactive isotopes, are present, though in extremely low concentrations, in ores and other natural substances, in products obtained synthetically, and in the solutions formed after processing raw materials. The decay these isotopes undergo is accompanied by nuclear radiation. Most natural radioisotopes are daughter isotopes, that is, products of the decay of ^{238}U, ^{235}U, and ^{232}Th.

Table: Concentration of daughter isotopes (in grams).

Parent isotope	^{210}po	^{223}Fr	^{22}Rn	^{227}Ac	^{226}Ra	^{228}Ra	^{228}Ac	^{231}Pa
^{238}U.......	7.6×10^{-11}		2.14×10^{-13}		3.4×10^{-7}			
^{235}U		1.3×10^{-15}		1×10^{-10}				5.6×10^{-5}
^{232}Th............						1.5×10^{-9}	5×10^{-14}	

Radioisotopes are also obtained artificially by irradiating various substances with nuclear particles. The yield is of the order of 10^{-8}–10^{-12} percent by weight. In many cases, hundreds, tens, or even just a few atoms of radioisotopes are present in many other atoms. (Only in the production of nuclear fuel is Pu obtained in significant quantities, though even here its concentration upon irradiation of U with neutrons is low.) It is therefore possible to separate radioactive elements and isotopes only from extremely dilute systems, and their weights in most cases cannot be determined. The physicochemical behavior of extremely dilute solutions is very complex. This behavior may be described by the laws of ideal solutions, though sometimes these laws are not obeyed because of secondary processes related to, for example, adsorption or radiolysis. General radio-chemistry includes the study of isotopic exchange, processes involving the distribution of trace amounts of radioisotopes between phases, processes of coprecipitation, adsorption, and extraction, the electrochemistry of radioactive elements, and the state of radioisotopes in extremely dilute systems—the dispersity of the elements (formation of radiocolloids) and the formation of complexes.

The chemistry of nuclear transformations includes the study of the reactions of atoms formed in nuclear transformations (hot atoms), the products of nuclear reactions, and the methods for obtaining, concentrating, and separating radioisotopes and their nuclear isomers. Also studied are the properties of radioactive substances and the transformations of these substances under the effect of their own radiation.

The chemistry of radioactive elements is the chemistry of the natural radioactive elements from Po to U (atomic numbers 84–92) and of the artificial elements Tc (atomic number 43), Pm (atomic number 61), Np (atomic number 94), and all subsequent elements up to atomic number 106. By convention, the chemistry and technology of nuclear fuel is included in this subdivision. Nuclear fuel entails the production and chemical separation of ^{239}Pu from irradiated uranium, of ^{233}U from thorium irradiated with neutrons, and of ^{235}U from a natural mixture of isotopes.

Applied radiochemistry is concerned with the development of methods for the synthesis of labeled compounds, the use of radioisotopes in chemistry and the chemical industry, and the use of nuclear radiation in chemical analysis, for example, in nuclear gamma-ray fluorescence spectroscopy.

The objects of radiochemical investigations are radioactive substances containing radioisotopes, many of which are characterized by a short lifetime and nuclear (radioactive) radiation, which necessitates the use of special methods of investigation.

Radioactive emissions permit the use of special radiometric methods in radiochemistry for measuring the quantity of a radioactive substance. These emissions, however, necessitate the use of special safety techniques because emissions in doses exceeding certain levels are harmful to human health. Since the methods for measuring radioactivity are of superior sensitivity, work can be done with a minimal amount of material, an amount that would not lend itself to study under any other methods. Using instruments common in radiochemical laboratories, it is possible to detect, for example, 10^{-10}—10^{-15} gram of ^{226}Ra, 10^{-17} gram of ^{32}P, or 10^{-17} gram of ^{222}Rn. By using especially sensitive methods for monitoring radioactive decay, it is possible to detect the presence of single atoms of a radioisotope and to establish the fact of their decay.

The development of radiochemistry into an independent branch of chemistry began at the end of the 19th century with the work of M. Skłodowska-Curie and P. Curie, who in 1898 discovered and isolated Ra and Po. Skłodowska-Curie was the first to use the methods of coprecipitation of trace amounts of radioactive elements from solution with large amounts of analogous elements. In 1911, F. Soddy defined radiochemistry as the science concerned with the study of the properties of the products of radioactive transformations and with the separation and identification of these products. Four periods can be discerned in the development of radiochemistry, each of which is related to progress in the study of radioactivity and nuclear physics.

The first period was characterized by the discovery of five natural radioactive elements—Po, Ra, Rn, Ac, and Pa—and a series of their isotopes (this became clear after the discovery by Soddy in 1913 of the existence of isotopes). As a result of the formulation by K. Fajans and Soddy of the displacement law, according to which a new element formed from a radioactive element is located in D. I. Mendeleev's periodic system either two places to the left of the parent element (α-decay) or one place to the right (β-decay), E. Rutherford and Soddy found a genetic relationship between all the discovered isotopes and determined the positions of these isotopes in the periodic system. In this period, an intense search was conducted for radioactive substances in nature, that is, radioactive minerals and waters. In Russia, A. P. Sokolov and other scientists studied the radioactivity of mineral waters and the atmosphere, P. P. Orlov initiated a study of the radioactivity of minerals, and V. I. Vernadskii did important work on the geochemistry of radioactive elements.

The second period was linked with the determination of a number of regularities in the behavior of radioisotopes in extremely dilute systems—solutions and gaseous systems—and with the discovery by G. de Hevesy and F. Paneth of isotopic exchange. Paneth and Fajans formulated laws of adsorption, and O. Hahn and V. G. Khlopin carried out systematic studies of the processes of coprecipitation and adsorption. As a result, Hahn formulated laws that qualitatively characterize these processes, Khlopin established a quantitative law of coprecipitation (Khlopin's law), and Khlopin's student A. P. Ratner developed a thermodynamic theory to account for the processes of distribution of a substance between a solid crystalline phase and a solution. Also in this period, another Soviet scientist, L. S. Kolovrat-Chervinskii, and then Hahn furthered the study of the emanations of solids containing radium isotopes, and B. A. Nikitin subsequently carried out extensive studies of the clathrate compounds of inert gases, as in the compounds of radon. In 1917, V. I. Spitsyn conducted a series of investigations on determining the solubility of a number of compounds of thorium through the use of radioactive tracers. The basis of such a method had been developed earlier by de Hevesy and Paneth.

During this period, Sktodowska-Curie and Paneth studied radioisotopes in extremely dilute solutions and the conditions necessary for the formation of radiocolloids.

The third period began after the discovery of artificial radioactivity by Jean Frédéric Joliot-Curie and his wife, Irène. In this period, the foundation was laid for methods of obtaining, concentrating, and separating artificial radioisotopes as a result of the work of E. Fermi on the action of neutrons on chemical elements, the discovery and study of the nuclear isomerism of artificial radioisotopes by I. V. Kurchatov and his coworkers, the discovery of the fission of uranium nuclei under the effect of neutrons by Hahn and the German scientist F. Strassmann, and the discovery of the Szilard-Chalmers reaction. The use of the cyclotron permitted E. Segrè and his coworkers to synthesize the new artificial elements Tc and At. Using radiometric methods, in conjunction with precise radiochemical methods for the separation of trace amounts of radioactive elements, M. Perey separated Fr (atomic number 87) from the decay products of Ac. From the mid-1930's, applied radiochemistry developed rapidly, and the method of radioactive (isotopic) tracers has received wide application.

The current, fourth period of the development of radiochemistry is linked with the use of powerful nuclear particle accelerators and nuclear reactors. This period has seen the synthesis and separation of artificial chemical elements—promethium by the American scientists J. Marinsky and L. Glendenin and the transuranium elements, with atomic numbers 93 to 105, by G. Seaborg, G. N. Flerov, and their respective co-workers. Methods for the production of nuclear fuel, the separation of Pu and fission products from the U irradiated in nuclear reactors, and the reprocessing of spent U from reactors have all been undergoing improvement. A number of important technological problems related to nuclear fuel have been solved. The chemistry of artificial, especially transuranium, and natural, especially U, Th, and Pa, radioactive elements has undergone extensive development, particularly in the chemistry of the complexes formed by these elements. The chemistry of new atomlike species—positronium, muonium, and mesic atoms—has also been established.

Extraction and chromatography have acquired special importance in radiochemistry. The use of radioactive tracer techniques is expanding in the study of the mechanisms and kinetics of chemical reactions, the structure of chemical compounds, and the phenomena of adsorption, coprecipitation, and catalysis. Tracer techniques are also used in the measurement of physicochemical constants and in the development of methods of radiometric analysis. Radiochemical methods are widely used in the solution of many problems in geochemistry and astro-chemistry and have been used in the search for mineral resources.

A new area in radiochemistry is the chemistry of processes that occur after a nuclear reaction, when the newly formed radioisotopes possess high energy. Studies are also under way on the products of nuclear transformations under the action of high-energy particles. Soviet scientists and scientists of a number of other countries are actively working in all these areas of radiochemistry. Radiochemistry is continuing to develop and to encompass new areas in the chemistry of radioactive substances.

Radiochemical methods tend to be labor intensive and generate liquid waste due to the chemical separations that are involved. Furthermore, detection limits tend to be high for long-lived isotopes using radiochemical methods.

Gamma-ray Emission

Gamma rays are produced by nuclear relaxations. Gamma-ray emission is the result of a nucleus in an excited state returning to the ground state in one or more quantized steps with the release of monoenergetic gamma rays. Gamma rays, except for their source, are indistinguishable from X-rays of the same energy. The gamma-ray emission spectrum is characteristic for each nucleus and is thus useful for identifying radioisotopes. Gamma radiation is highly penetrating. Gamma rays in interacting with matter lose energy by three mechanisms; the one that predominates depends upon the energy of the gamma-ray photon. With low-energy gamma radiation, the photoelectric effect predominates. In this interaction, the gamma-ray photon disappears after rejecting an electron from an atomic orbital of the target atom. The photon energy is totally consumed in overcoming the binding energy of the electron and in overcoming the binding energy of the electron and in imparting kinetic energy to the ejected electron. Relatively energetic gamma interacts via the compton effect. In this, an electron is also ejected from an atom but acquires only a part of the photon energy. The photon now with diminished energy, recoiled from the electron and goes on to further Compton or photoelectric interactions. If the gamma-ray photon possesses sufficiently high energy (at least 1.02 MeV), pair production can occur. In this scenario the photon is totally absorbed in creating a positron and an electron in the field surrounding a nucleus.

Paul Villard, a French physicist, is credited with discovering gamma rays. Villard recognized them as different from X-rays because the gamma rays had a much greater penetrating depth. It wasn't until 1914 that Rutherford showed that they were a form of light with a much shorter wavelength than X-rays.

Gamma rays are more dangerous than radio waves. This is due to the fact that light can be thought of as particles (photons) as well as electromagnetic waves. A radio photon doesn't have much energy and doesn't travel through matter well (that's why you don't pick up radio well in a tunnel). A gamma-ray photon has enough energy to damage atoms in your body and make them radioactive, and gamma rays can easily penetrate into your body. It's like the difference between getting hit by sand or a bullet. It takes a lot of sand to do any damage, but only one bullet.

X-ray Emission

X-Ray emission is formed from electronic transitions in which outer electrons fill the vacancies created by the nuclear process. One of the processes is electron capture. A second process which may lead to X-rays is internal conversion, a type of nuclear process that is an alternative to gamma-ray emission. In this process, an electromagnetic interaction between the excited nucleus and an extranuclear electron results in the ejection of an orbital electron with a kinetic energy equal to the difference between the energy of the electron. The emission of this so-called internal conversion electron leaves a vacancy in the K, L or higher orbitals. X-rays are emitted as the orbital is filled by an electronic transition.

On the earth, we generate the energetic particles in analytical instruments such as the scanning electron microscope (called an electron microprobe) and measure the energies of the X-rays emitted from the sample. The general concept: electrons are knocked out of occupied levels by energetic X-rays, electrons, or ions (either protons - H^+ - or alpha particles - He^{++}), electrons make transitions from filled to empty states and emit an X-ray. The energy of the emitted characteristic X-ray identifies the atom.

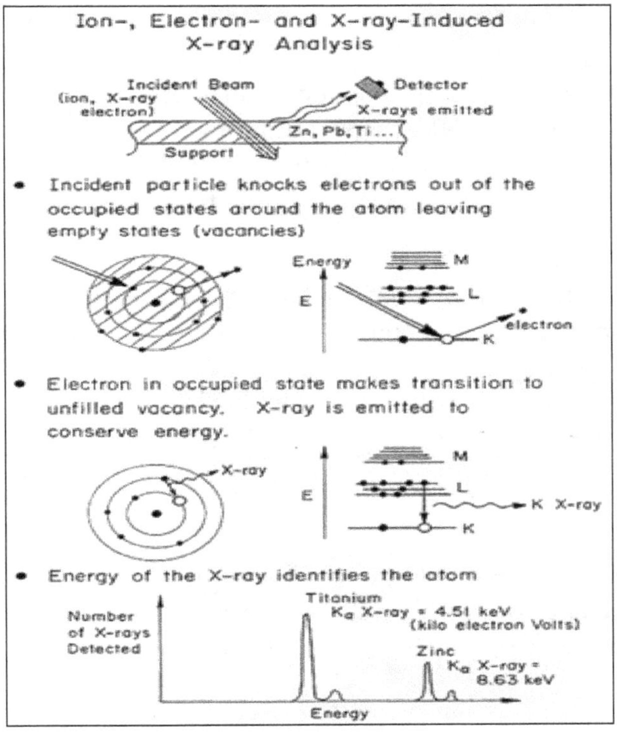

Radioactive Decay Rates

Radioactive decay is a completely random process. Although no predictions can be made concerning the lifetime of an individual nucleus, the behavior of a large ensemble of like nuclei can be described by the first order rate expression.

$$-\frac{dN}{dt} = \lambda N$$

Where N is the number of radioactive nuclei of a particular kind in the sample at time t and λ is the decay constant for the particular radioisotope. Over this interval this expression can be rearranged to,

$$N = N_0 e^{-kt}$$

The half-life, $t_{½}$ is,

$$t_{½} = \frac{\ln 2}{\lambda} \approx \frac{0.693}{\lambda}$$

The half-lives of radioactive species range from small fractions of a second to many billions of years. The activity A or a radionuclide is defined by its disintegration rate.

$$A = -\frac{dN}{dt} = \lambda N$$

Activity is given in units of s^{-1}. The becquerel (Bq) corresponds to 1 decay per second (1 Bq = 1 s^{-1}). An older, but still widely used unit of activity is the Curie (Ci), which was originally defined as the activity of 1 g of radium-226. One curie is equal to 3.70×10^{10} Bq. In analytical radiochemistry, activities of analytes range from a nanocurie to less to a few microcuries.

In the laboratory, solution activities are seldom measured because detection efficiencies are seldom 100%. Instead, the counting rate, R is employed, where,

$$R = cA$$

Substituting this into the activity expression yields,

$$R = c\lambda N$$

Here, c is constant called the detection coefficient, which depends upon the nature of the detector, the efficiency of counting disintegrations, and the geometric arrangement of sample and detector. The decay law given then can be written as,

$$R = R_0 e^{-kt}$$

Radioactivity is measured by means of a detector that produces a pulse of electricity for each atom undergoing decay. Quantitative information about decay rates is obtained by counting these pulses for a specific period. Shown below is a table with decay data obtained by successive one-minute counts.

Minutes	Counts	Minutes	Counts
1	180	7	168
2	187	8	170
3	166	9	173

4	173	10	132
5	170	11	154
6	164	12	167

Total counts = 2004, Average counts/min = 167.

Radioactive decay is random. In order to accurately describe radioactive behavior, it is necessary to assume a Poisson distribution, which is given by the equation,

$$y = \frac{\mu^{x_i}}{x_i\,!}e^{-\mu}$$

where y is the frequency of occurrence of a given count x_i and μ is the mean for a large set of counting data. The data plotted in the figure below, shows the deviation (x_i - μ) from the true average count that would be expected if 1000 replicate observations were made on the same sample. Curve A gives the distribution for a substance for which the true average count μ for a selected period is 5. Curve B and C correspond to samples having the true means of 15 and 35. The absolution deviation becomes greater with increases in μ, but the relative deviations become smaller. Note that for the two smaller numbers of counts, the distribution is not symmetric around the average. This lack of symmetry is a consequence of the fact that a negative count is impossible, while a finite probability exists that a given count will exceed the average by severalfold.

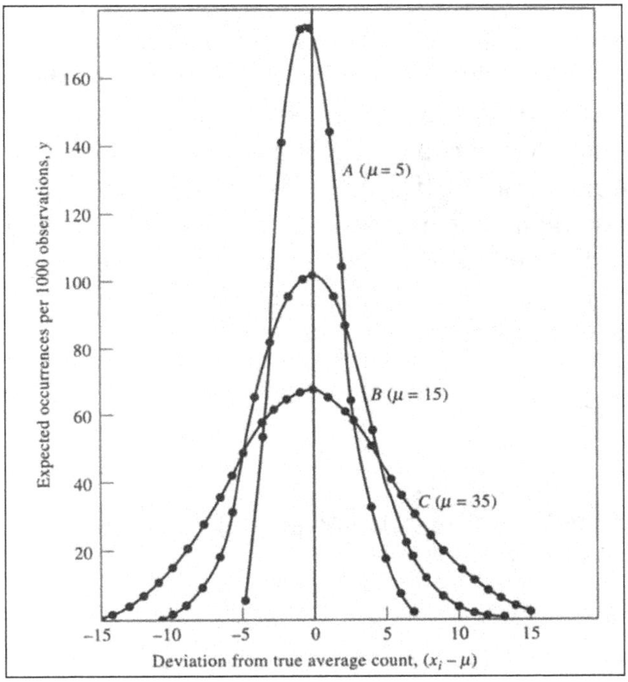

The confidence interval for a measurement was defined as the limits around a measured quantity within the true mean can be expected to fall with a stated probability.

When the measured standard deviation is believed to be a good approximation of the true standard deviation ($S \to \sigma$), the confidence limit CL is,

$$\text{CL for } \mu = \bar{x} \pm z\sigma$$

For counting rates, this equation takes the following form,

$$\text{CL for } R = R \pm z\sigma_R$$

where *z* is dependent upon the desired level of confidence. The relative uncertainties in counting are shown below.

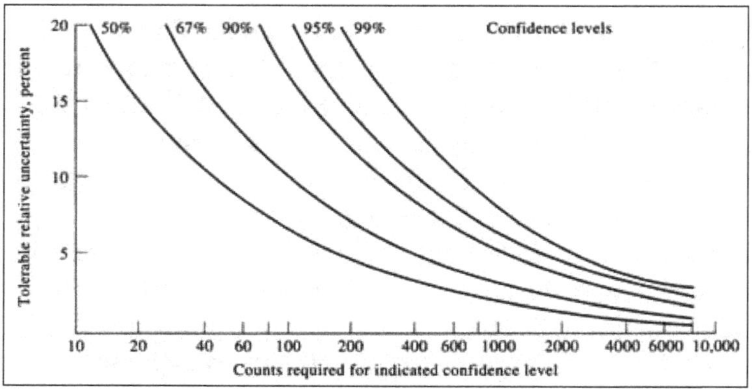

The counts recorded in a radiochemical analysis include a contribution from sources other than the sample. Background activity can be traced to the existence of minute quantities of radon isotopes in the atmosphere, to the materials used in construction of the laboratory, to accidental contamination within the laboratory, to cosmic radiation, and to the release of radioactive materials into the Earth's atmosphere. In order to obtain a true assay, then, it is necessary to correct the total count for background. The counting period required to establish the background correction frequently differs from that for the sample, as a result, it is more convenient to employ counting rates. The standard deviation of the corrected counting rate is,

$$\sigma_{R_C} = \sqrt{\frac{R_x}{t_x} + \frac{R_b}{t_b}}$$

Where R_c is the corrected counting rate and R_x and R_b are the rates for the sample and the background, respectively.

Instrumentation

Radiation from radioactive sources can be detected and measured in essentially the same way as X-radiation. Gas-filled transducers, scintillation counters and semiconductor detectors are all sensitive to alpha and beta particles and gamma rays because absorption of these particles induces ionization or photoelectrons, which can in turn produce thousands of ion pairs. Alpha-emitting samples are generally counted as thin deposits prepared by electrodeposition or by distillation and condensation. These deposits are then sealed with think windows and counted in windowless gas-flow proportional counters or ionization chambers. Liquid scintillation counting is becoming more important for counting alpha emitters because of the ease of sample preparation and the higher sensitivity for alpha-particle detection.

For beta sources having energies greater than about 0.2 MeV, a uniform layer of the sample is counted with a thin-windowed Geiger or proportional tube counter. For low-energy beta emitter, such as carbon-14, sulfur-35 and tritium, liquid scintillation counters are preferable. In a liquid scintillation counter a vial containing the solution is placed between two photomultiplier tubes housed in a light-tight container. The output from the two transducers is fed into a coincidence counter, an electronic device that only records a count only when pules from the two transducers arrive at same. The coincidence counter reduces the background noise from the detector and amplifiers because of the low probability that such noise will affect both systems simultaneously.

Gamma radiation is detected and measured by the methods used for the detection for X-radiation. Interference from alpha- and beta- radiation is readily avoided by filtering the radiation with a thin film piece of aluminum or Mylar. Gamma-ray spectrometer is similar to the pulse-height analyzer. The figure below shows a gamma-ray reference spectrum obtained with a 4000 channel analyzer. The characteristic peaks for the various elements are superimposed on the continuum that arises primarily from the Compton effect.

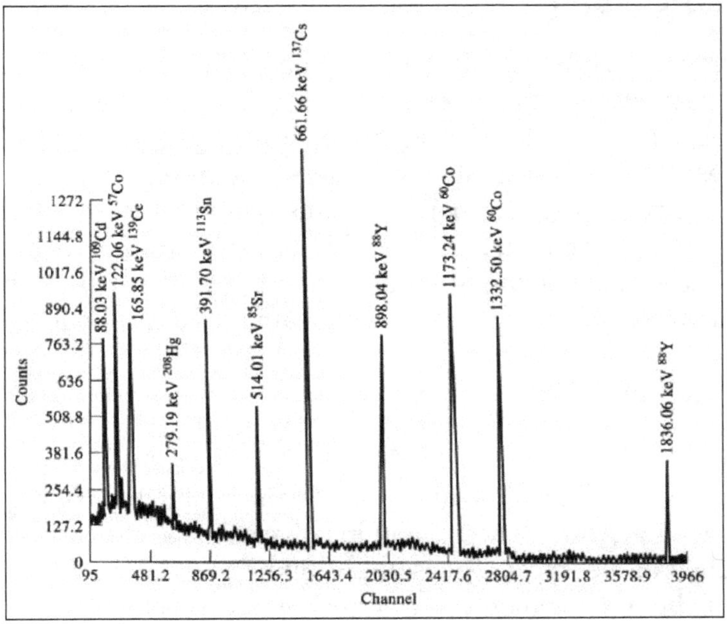

The schematic below shows a well-type scintillation counter that is used for gamma-ray counting.

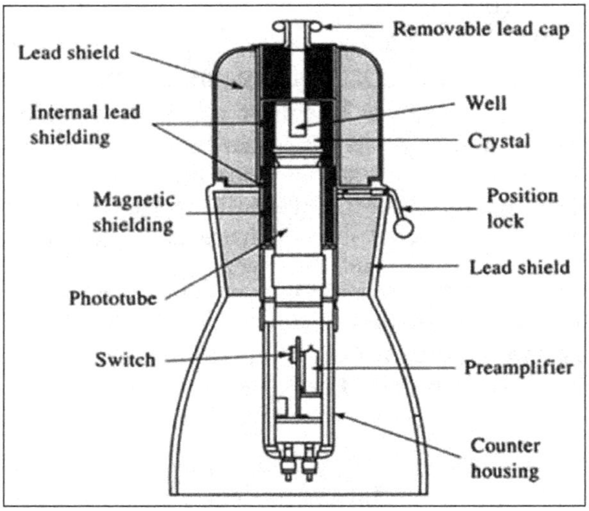

Isotope Dilution Method

Isotope dilution methods predate activation procedure and are still being extensively applied to problems in all branches of chemistry. Stable and radioactive isotopes are employed in the isotope dilution technique. The isotope dilution method is more convenient because of the ease with which the concentration of the isotope can be determined.

Principles of the Isotope Dilution Procedure

Isotope dilution methods require the preparation of a quantity of the analyte in a radioactive form. A known weight of the isotopically labeled species is then mixed with the sample to be analyzed. After treating to assure homogeneity between the active and nonactive species, a part of the analyte is isolated chemically in the form of a purified compound. By counting a weighed portion of this product, the extent of dilution of the active material can be calculated and related to the amount of nonactive substance in the original sample. The following equation relates the activity of the isolated and purified mixture of analyte and tracer to the original amount of the analyte,

$$W_x = \frac{R_t}{R_m} W_m - W_t$$

where, W_x is the grams of tracer having a counting rate of R_t (cpm), W_m is the grams of isolated and purified mixture of active and inactive species with a counting rate of R_m, and W_t is the weight of the tracer. The isotopic dilution technique has been used for the determination of about 30 elements in a variety of matrix materials.

Modern uses of Radioactive Isotopes

Smoke Detectors and Americium - 241

How many of us have smoke detectors in our house? Chances are that a great number of homes have had one or more of these devices installed as an early warning system in case of fire. What most consumers don't know is that many of these units contain a small amount of americium-241. By utilizing the radioactive properties of this material, smoke from a fire can be detected at a very early stage. This early warning capability has saved many lives. In fact, studies have shown that 80% of fire injuries and 80% of fire fatalities occur in homes without smoke detectors.

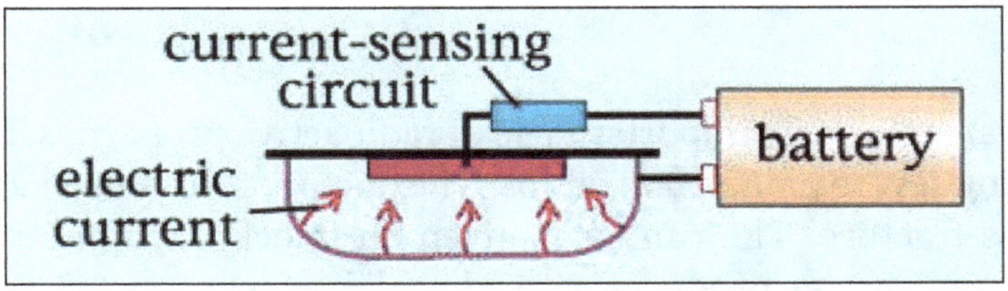

Agricultural Applications - Radioactive Tracers

Radioisotopes can be used to help understand chemical and biological processes in plants. This is true for two reasons: 1) radioisotopes are chemically identical with other isotopes of the same element and will be substituted in chemical reactions and 2) radioactive forms of the element can be easily detected with a Geiger counter or other such device.

A solution of phosphate, containing radioactive phosphorus-32, is injected into the root system of a plant. Since phosphorus-32 behaves identically to that of phosphorus-31, the more common and non-radioactive form of the element, the plant uses it in the same way. A Geiger counter is then used to detect the movement of the radioactive phosphorus-32 throughout the plant. This information helps scientists understand the detailed mechanism of how plants utilized phosphorus to grow and reproduce.

Archaeological Dating

Significant progress has been made in this field of study since the discovery of radioactivity and its properties. One application is carbon-14 dating. Recalling that all biologic organisms contain a given concentration of carbon-14, we can use this information to help solve questions about when the organism died. It works like this: when an organism dies it has a specific ratio by mass of carbon-14 to carbon-12 incorporated in the cells of its body. (The same ratio as in the atmosphere.) At the moment of death, no new carbon-14 containing molecules are metabolized; therefore the ratio is at a maximum. After death, the carbon-14 to carbon-12 ratio begins to decrease because carbon-14 is decaying away at a constant and predictable rate. Remembering that the half-life of carbon-14 is 5700 years, then after 5700 years half as much carbon-14 remains within the organism.

Medical Uses

Bone imaging is an extremely important use of radioactive properties. Supposed a runner is experiencing severe pain in both shins. The doctor decides to check to see if either tibia has a stress fracture. The runner is given an injection containing 99Tcm. This radioisotope is a gamma ray producer with a half-life of 6 hours.

After a several hour wait, the patient undergoes bone imaging. At this point, any area of the body that is undergoing unusually high bone growth will show up as a stronger image on the screen. Therefore if the runner has a stress fracture, it will show up on the bone-imaging scan. This technique is also good for arthritic patients, bone abnormalities and various other diagnostics.

Radioactive Isotope

Radioactive isotope, also called radioisotope, radionuclide, or radioactive nuclide is any of several species of the same chemical element with different masses whose nuclei are unstable and dissipate excess energy by spontaneously emitting radiation in the form of alpha, beta, and gamma rays.

A brief treatment of radioactive isotopes follows.

Every chemical element has one or more radioactive isotopes. For example, hydrogen, the lightest element, has three isotopes with mass numbers 1, 2, and 3. Only hydrogen-3 (tritium), however, is a radioactive isotope, the other two being stable. More than 1,000 radioactive isotopes of the various elements are known. Approximately 50 of these are found in nature; the rest are produced artificially as the direct products of nuclear reactions or indirectly as the radioactive descendants of these products.

Radioactive isotopes have many useful applications. In medicine, for example, cobalt-60 is extensively employed as a radiation source to arrest the development of cancer. Other radioactive isotopes are used as tracers for diagnostic purposes as well as in research on metabolic processes. When a radioactive isotope is added in small amounts to comparatively large quantities of the stable element, it behaves exactly the same as the ordinary isotope chemically; it can, however, be traced with a Geiger counter or other detection device. Iodine-131 has proved effective in treating hyperthyroidism. Another medically important radioactive isotope is carbon-14, which is used in a breath test to detect the ulcer-causing bacteria Heliobacter pylori.

In industry, radioactive isotopes of various kinds are used for measuring the thickness of metal or plastic sheets; their precise thickness is indicated by the strength of the radiations that penetrate the material being inspected. They also may be employed in place of large X-ray machines to examine manufactured metal parts for structural defects. Other significant applications include the use of radioactive isotopes as compact sources of electrical power—e.g., plutonium-238 in spacecraft. In such cases, the heat produced in the decay of the radioactive isotope is converted into electricity by means of thermoelectric junction circuits or related devices.

Radioactive Decay

Radioactive decay or radioactivity is the property exhibited, by certain types of matter of emitting energy and subatomic particles spontaneously. It is, in essence, an attribute of individual atomic nuclei.

An unstable nucleus will decompose spontaneously, or decay, into a more stable configuration but will do so only in a few specific ways by emitting certain particles or certain forms of electromagnetic energy. Radioactive decay is a property of several naturally occurring elements as well as of artificially produced isotopes of the elements. The rate at which a radioactive element decays is expressed in terms of its half-life; i.e., the time required for one-half of any given quantity of the isotope to decay. Half-lives range from more than 1,000,000,000 years for some nuclei to less than 10^{-9} second. The product of a radioactive decay process—called the daughter of the parent isotope—may itself be unstable, in which case it, too, will decay. The process continues until a stable nuclide has been formed.

Nature of Radioactive Emissions

The emissions of the most common forms of spontaneous radioactive decay are the alpha (α) particle, the beta (β) particle, the gamma (γ) ray, and the neutrino. The alpha particle is actually the nucleus of a helium-4 atom, with two positive charges $^{4}_{2}He$. Such charged atoms are called ions. The neutral helium atom has two electrons outside its nucleus balancing these two charges. Beta particles may be negatively charged (beta minus, symbol e^-), or positively charged (beta plus, symbol e^+). The beta minus (β^-) particle is actually an electron created in the nucleus during beta decay without any relationship to the orbital electron cloud of the atom. The beta plus particle, also called the positron, is the antiparticle of the electron; when brought together, two such particles will mutually annihilate each other. Gamma rays are electromagnetic radiations such as radio waves, light, and X-rays. Beta radioactivity also produces the neutrino and antineutrino, particles that have no charge and very little mass, symbolized by v and $\bar{v}$, respectively.

In the less common forms of radioactivity, fission fragments, neutrons, or protons may be emitted. Fission fragments are themselves complex nuclei with usually between one-third and two-thirds the charge Z and mass A of the parent nucleus. Neutrons and protons are, of course, the basic building blocks of complex nuclei, having approximately unit mass on the atomic scale and having zero charge or unit positive charge, respectively. The neutron cannot long exist in the free state. It is rapidly captured by nuclei in matter; otherwise, in free space it will undergo beta-minus decay to a proton, an electron, and an antineutrino with a half-life of 12.8 minutes. The proton is the nucleus of ordinary hydrogen and is stable.

Occurrence of Radioactivity

Some species of radioactivity occur naturally on Earth. A few species have half-lives comparable to the age of the elements (about 6×10^9 years), so that they have not decayed away after their formation in stars. Notable among these are uranium-238, uranium-235, and thorium-232. Also, there is potassium-40, the chief source of irradiation of the body through its presence in potassium of tissue. Of lesser significance are the beta emitters vanadium-50, rubidium-87, indium-115, tellurium-123, lanthanum-138, lutetium-176, and rhenium-187, and the alpha emitters cerium-142, neodymium-144, samarium-147, gadolinium-152, dysprosium-156, hafnium-174, platinum-190, and lead-204. Besides these approximately 10^9-year species, there are the shorter-lived daughter activities fed by one or another of the above species; e.g., by various nuclei of the elements between lead ($Z = 82$) and thorium ($Z = 90$).

Another category of natural radioactivity includes species produced in the upper atmosphere by cosmic ray bombardment. Notable are 5,720-year carbon-14 and 12.3-year tritium (hydrogen-3), 53-day beryllium-7, and 2,700,000-year beryllium-10. Meteorites are found to contain additional small amounts of radioactivity, the result of cosmic ray bombardments during their history outside the Earth's atmospheric shield. Activities as short-lived as 35-day argon-37 have been measured in fresh falls of meteorites. Nuclear explosions since 1945 have injected additional radioactivities into the environment, consisting of both nuclear fission products and secondary products formed by the action of neutrons from nuclear weapons on surrounding matter.

The fission products encompass most of the known beta emitters in the mass region 75–160. They are formed in varying yields, rising to maxima of about 7 percent per fission in the mass region 92–102 (light peak of the fission yield versus atomic mass curve) and 134–144 (heavy peak). Two kinds of delayed hazards caused by radioactivity are recognized. First, the general radiation level is raised by fallout settling to Earth. Protection can be provided by concrete or earth shielding until the activity has decayed to a sufficiently low level. Second, ingestion or inhalation of even low levels of certain radioactive species can pose a special hazard, depending on the half-life, nature of radiations, and chemical behaviour within the body.

Nuclear reactors also produce fission products but under conditions in which the activities may be contained. Containment and waste-disposal practices should keep the activities confined and eliminate the possibility of leaching into groundwaters for times that are long compared to the half-lives. A great advantage of thermonuclear fusion power over fission power, if it can be practically realized, is not only that its fuel reserves, heavy hydrogen and lithium, are vastly greater than uranium, but also that the generation of radioactive fission product wastes can be largely avoided. In this connection, it may be noted that a major source of heat in the interior of both the Earth and the Moon is provided by radioactive decay. Theories about the formation and evolution of the Earth, Moon, and other planets must take into account these large heat production sources.

Desired radioactivities other than natural activities and fission products may be produced either by irradiation of certain selected target materials by reactor neutrons or by charged particle beams or gamma ray beams of accelerators.

Energetics and Kinetics of Radioactivity

Energy Release in Radioactive Transitions

Consideration of the energy release of various radioactive transitions leads to the fundamental question of nuclear binding energies and stabilities. A much-used method of displaying nuclear-stability relationships is an isotope chart, those positions on the same horizontal row corresponding to a given proton number (Z) and those on the same vertical column to a given neutron number (N). The irregular bold line surrounds the region of presently known nuclei. The area encompassed by this is often referred to as the valley of stability because the chart may be considered a map of a binding energy surface, the lowest areas of which are the most stable. The most tightly bound nuclei of all are the abundant iron and nickel isotopes. Near the region of the valley containing the heaviest nuclei (largest mass number A; i.e., largest number of nucleons, $N + Z$), the processes of alpha decay and spontaneous fission are most prevalent; both these processes relieve the energetically unfavourable concentration of positive charge in the heavy nuclei.

Along the region that borders on the valley of stability on the upper left-hand side are the positron-emitting and electron-capturing radioactive nuclei, with the energy release and decay rates increasing the farther away the nucleus is from the stability line. Along the lower right-hand border region, beta-minus decay is the predominant process, with energy release and decay rates increasing the farther the nucleus is from the stability line.

The grid lines of the graph are at the nucleon numbers corresponding to extra stability, the "magic numbers". The circles labeled "deformed regions" enclose regions in which nuclei should exhibit cigar shapes; elsewhere the nuclei are spherical. Outside the dashed lines nuclei would be unbound with respect to neutron or proton loss and would be exceedingly short-lived (less than 10^{-19} second).

Calculation and Measurement of Energy

By the method of closed energy cycles, it is possible to use measured radioactive-energy-release (Q) values for alpha and beta decay to calculate the energy release for unmeasured transitions. An illustration is provided by the cycle of four nuclei below.

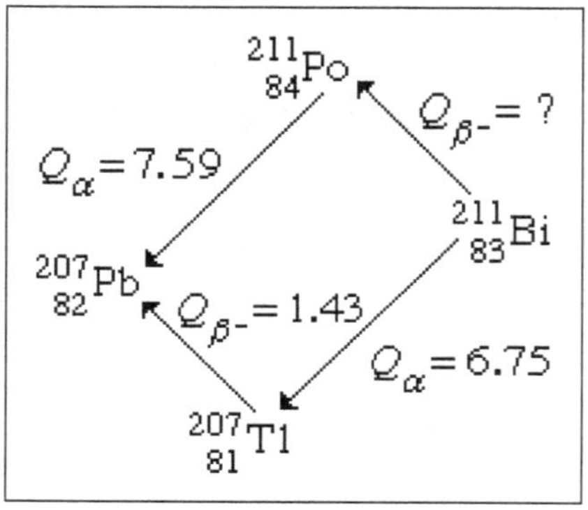

In this cycle, energies from two of the alpha decays and one beta decay are measurable. The unmeasured beta-decay energy for bismuth-211, $Q_{\beta-}$(Bi), is readily calculated because conservation of energy requires the sum of Q values around the cycle to be zero. Thus, $Q_{\beta-}$(Bi) + 7.59 − 1.43 − 6.75 = 0. Solving this equation gives $Q_{\beta-}$(Bi) = 0.59 MeV. This calculation by closed energy cycles can be extended from stable lead-207 back up the chain of alpha and beta decays to its natural precursor uranium-235 and beyond. In this manner the nuclear binding energies of a series of nuclei can be linked together. Because alpha decay decreases the mass number A by 4, and beta decay does not change A, closed α–β-cycle calculations based on lead-207 can link up only those nuclei with mass numbers of the general type $A = 4n + 3$, in which n is an integer. Another, the $4n$ series, has as its natural precursor thorium-232 and its stable end product lead-208. Another, the $4n + 2$ series, has uranium-238 as its natural precursor and lead-206 as its end product.

In early research on natural radioactivity, the classification of isotopes into the series cited above was of great significance because they were identified and studied as families. Newly discovered radioactivities were given symbols relating them to the family and order of occurrence therein. Thus, thorium-234 was known as UX_1, the isomers of protactinium-234 as UX_2 and UZ, uranium-234 as U_{II}, and so forth. These original symbols and names are occasionally encountered in more recent literature but are mainly of historical interest. The remaining $4n + 1$ series is not naturally occurring but comprises well-known artificial activities decaying down to stable thallium-205.

To extend the knowledge of nuclear binding energies, it is clearly necessary to make measurements to supplement the radioactive-decay energy cycles. In part, this extension can be made by measurement of Q values of artificial nuclear reactions. For example, the neutron-binding energies of the lead isotopes needed to link the energies of the four radioactive families together can be measured by determining the threshold gamma-ray energy to remove a neutron (photonuclear reaction); or the energies of incoming deuteron and outgoing proton in the reaction can be measured to provide this information.

Further extensions of nuclear-binding-energy measurements rely on precision mass spectroscopy. By ionizing, accelerating, and magnetically deflecting various nuclides, their masses can be measured with great precision. A precise measurement of the masses of atoms involved in radioactive decay is equivalent to direct measurement of the energy release in the decay process. The atomic mass of naturally occurring but radioactive potassium-40 is measured to be 39.964008 amu. Potassium-40 decays predominantly by β-emission to calcium-40, having a measured mass 39.962589. Through Einstein's equation, energy is equal to mass (m) times velocity of light (c) squared, or $E = mc^2$, the energy release (Q) and the mass difference, Δm, are related, the conversion factor being one amu, equal to 931.478 MeV. Thus, the excess mass of potassium-40 over calcium-40 appears as the total energy release Q_β in the radioactive decay $Q_{\beta-} = (39.964008 - 39.962589) \times 931.478$ MeV = 1.31 MeV. The other neighbouring isobar (same mass number, different atomic number) to argon-40 is also of lower mass, 39.962384, than potassium-40. This mass difference converted to energy units gives an energy release of 1.5 MeV, this being the energy release for EC decay to argon-40. The maximum energy release for positron emission is always less than that for electron capture by twice the rest mass energy of an electron ($2m_0c^2 = 1.022$ MeV); thus, the maximum positron energy for this reaction is 1.5 – 1.02, or 0.48 MeV.

To connect alpha-decay energies and nuclear mass differences requires a precise knowledge of the alpha-particle (helium-4) atomic mass. The mass of the parent minus the sum of the masses of the decay products gives the energy release. Thus, for alpha decay of plutonium-239 to uranium-235 and helium-4 the calculation goes as follows:

$$\begin{array}{lr} M(^{239}\mathrm{Pu}) & 239.05216 \\ -M(^{235}\mathrm{U}) & -235.04393 \\ -M(^{4}\mathrm{He}) & -\quad 4.00260 \\ \hline & 0.00563 \times 931.478 \end{array}$$

$$Q_a = 5.24\,\mathrm{MeV}$$

By combining radioactive-decay-energy information with nuclear-reaction Q values and precision mass spectroscopy, extensive tables of nuclear masses have been prepared. From them the Q values of unmeasured reactions or decay may be calculated.

Alternative to the full mass, the atomic masses may be expressed as mass defect, symbolized by the Greek letter delta, Δ (the difference between the exact mass M and the integer A, the mass number), either in energy units or atomic mass units.

Absolute Nuclear Binding Energy

The absolute nuclear binding energy is the hypothetical energy release if a given nuclide were synthesized from Z separate hydrogen atoms and N (equal to $A - Z$) separate neutrons. An example is the calculation giving the absolute binding energy of the stablest of all nuclei, iron-56:

$$\begin{array}{ll} 26\times M(^{1}\text{H}) & 26\times 1.007825 = 26.20345 \\ 30\times M(n) & 30\times 1.008665 = 30.25995 \\ M(^{56}\text{Fe}) & \underline{-55.93493} \\ & \text{binding energy} = 0.52847\times 931.478 = \\ & \qquad 492.58\,\text{MeV} \end{array}$$

$$\text{average binding energy per nucleon of}\ ^{56}\text{Fe} = 492.58/56 = 8.796\,\text{MeV}$$

A general survey of the average binding energy per nucleon (for nuclei of all elements grouped according to ascending mass) shows a maximum at iron-56 falling off gradually on both sides to about 7 MeV at helium-4 and to about 7.4 MeV for the most massive nuclei known. Most of the naturally occurring nuclei are thus not stable in an absolute nuclear sense. Nuclei heavier than iron would gain energy by degrading into nuclear products closer to iron, but it is only for the elements of greatest mass that the rates of degradation processes such as alpha decay and spontaneous fission attain observable rates. In a similar manner, nuclear energy is to be gained by fusion of most elements lighter than iron. The coulombic repulsion between nuclei, however, keeps the rates of fusion reactions unobservably low unless the nuclei are subjected to temperatures of greater than 10^7 K. Only in the hot cores of the Sun and other stars or in thermonuclear bombs or controlled fusion plasmas are these temperatures attained and nuclear-fusion energy released.

Nuclear Models

The Liquid-drop Model

The average behaviour of the nuclear binding energy can be understood with the model of a charged liquid drop. In this model, the aggregate of nucleons has the same properties of a liquid drop, such as surface tension, cohesion, and deformation. There is a dominant attractive-binding-energy term proportional to the number of nucleons A. From this must be subtracted a surface-energy term proportional to surface area and a coulombic repulsion energy proportional to the square of the number of protons and inversely proportional to the nuclear radius. Furthermore, there is a symmetry-energy term of quantum-mechanical origin favouring equal numbers of protons and neutrons. Finally, there is a pairing term that gives slight extra binding to nuclei with even numbers of neutrons or protons.

The pairing-energy term accounts for the great rarity of odd–odd nuclei (the terms odd–odd, even–even, even–odd, and odd–even refer to the evenness or oddness of proton number, Z, and neutron number, N, respectively) that are stable against beta decay. The sole examples are deuterium, lithium-6, boron-10, and nitrogen-14. A few other odd–odd nuclei, such as potassium-40, occur in nature, but they are unstable with respect to beta decay. Furthermore, the pairing-energy term makes for the larger number of stable isotopes of even-Z elements, compared to odd-Z, and for the lack of stable isotopes altogether in element 43, technetium, and element 61, promethium.

The beta-decay energies of so-called mirror nuclei afford one means of estimating nuclear sizes. For example, the neon and fluorine nuclei, ${}^{19}_{10}Ne_9$ and ${}^{19}_{9}F_{10}$, are mirror nuclei because the proton and neutron numbers of one of them equal the respective neutron and proton numbers of the other. Thus, all binding-energy terms are the same in each except for the coulombic term, which is inversely proportional to the nuclear radius. Such calculations along with more direct determinations by high-energy electron scattering and energy measurements of X-rays from muonic atoms (hydrogen atoms in which the electrons are replaced by negative muons) establish the nuclear charge as roughly uniformly distributed in a sphere of radius $1.2\,A^{1/3} \times 10^{-13}$ centimetre. That the radius is proportional to the cube root of the mass number has the great significance that the average density of all nuclei is nearly constant.

Careful examination of nuclear-binding energies reveals periodic deviations from the smooth average behaviour of the charged-liquid-drop model. An extra binding energy arises in the neighbourhood of certain numbers of neutrons or protons, the so-called magic numbers (2, 8, 20, 28, 50, 82, and 126). Nuclei such as ${}^{4}_{2}He_2$, ${}^{16}_{8}O_8$, ${}^{40}_{20}Ca_{20}$, ${}^{48}_{20}Ca_{28}$, and ${}^{208}_{82}Pb_{126}$ are especially stable species, doubly magic, in view of their having both proton and neutron numbers magic. These doubly magic nuclei are situated at the intersections of grid lines.

The Shell Model

The overall trends of nuclear binding energies were described in terms of a charged-liquid-drop model. Yet there were noted periodic binding-energy irregularities at the magic numbers. The periodic occurrence of magic numbers of extra stability is strongly analogous to the extra electronic stabilities occurring at the atomic numbers of the noble-gas atoms. The explanations of these stabilities are quite analogous in atomic and nuclear cases as arising from filling of particles into quantized orbitals of motion. The completion of filling of a shell of orbitals is accompanied by an extra stability. The nuclear model accounting for the magic numbers is, as previously noted, the shell model. In its simplest form, this model can account for the occurrence of spin zero for all even–even nuclear ground states; the nucleons fill pairwise into orbitals with angular momenta canceling. The shell model also readily accounts for the observed nuclear spins of the odd-mass nuclei adjacent to doubly magic nuclei, such as $^{208}_{82}Pb$. Here, the spins of 1/2 for neighbouring $^{207}_{81}Tl$ and $^{207}_{82}Pb$ are accounted for by having all nucleons fill pairwise into the lowest energy orbits and putting the odd nucleon into the last available orbital before reaching the doubly magic configuration (the Pauli exclusion principle dictates that no more than two nucleons may occupy a given orbital, and their spins must be oppositely directed); calculations show the last available orbitals below lead-208 to have angular momentum 1/2. Likewise, the spins of 9/2 for $^{209}_{82}Pb$ and $^{209}_{83}Bi$ are understandable because spin-9/2 orbitals are the next available orbitals beyond doubly magic lead-208. Even the associated magnetization, as expressed by the magnetic dipole moment, is rather well explained by the simple spherical-shell model.

The orbitals of the spherical-shell model are labeled in a notation close to that for electronic orbitals in atoms. The orbital configuration of calcium-40 has protons and neutrons filling the following orbitals: $1s_{1/2}$, $1p_{3/2}$, $1p_{1/2}$, $1d_{5/2}$, and $1d_{3/2}$. The letter denotes the orbital angular momentum in usual spectroscopic notation, in which the letters *s*, *p*, *d*, *f*, *g*, *h*, *i*, etc., represent integer values of *l* running from zero for *s* through six for *i*. The fractional subscript gives the total angular momentum *j* with values of $l + 1/2$ and $l - 1/2$ allowed, as the intrinsic spin of a nucleon is 1/2. The first integer is a radial quantum number taking successive values 1, 2, 3, etc., for successively higher energy values of an orbital of given *l* and *j*. Each orbital can accommodate a maximum of $2j + 1$ nucleons. The exact order of various orbitals within a shell differs somewhat for neutrons and protons. The parity associated with an orbital is even (+) if *l* is even (*s*, *d*, *g*, *i*) and odd (–) if *l* is odd (*p*, *f*, *h*).

Table: Spherical-shell-model orbitals.

Shell Closure Number	
2	$1s_{1/2}$
8	$1p_{3/2}$, 1p
20	$1d_{5/2}$, $2s_{1/2}$, $1d_{3/2}$
28	$1f_{7/2}$
50	$2p_{3/2}$, $1f_{5/2}$, $2p_{1/2}$, $1g_{9/2}$
82	$1g_{7/2}$, $2d_{5/2}$, $1h_{11/2}$, $2d_{3/2}$, $3s_{1/2}$
126	$2f_{7/2}$, $1h_{9/2}$, $1i_{13/2}$, $3p_{3/2}$, $2f_{5/2}$, $3p_{1/2}$
184 (?)	$2g_{9/2}$, $1i_{11/2}$, $1j_{15/2}$, $3d_{5/2}$, $2g_{7/2}$, $4s_{1/2}$, $3d_{3/2}$

An example of a spherical-shell-model interpretation is provided by the beta-decay scheme of 2.2-minute thallium-209 shown below, in which spin and parity are given for each state. The ground and lowest excited states of lead-209 are to be associated with occupation by the 127th neutron of the lowest available orbitals above the closed shell of 126. From the last line of the table, it is to be noted that there are available $g_{9/2}$, $d_{5/2}$, and $s_{1/2}$ orbitals with which to explain the ground and first two excited states.

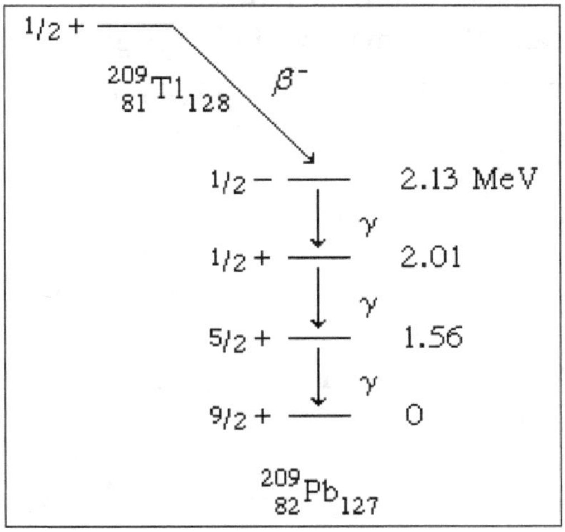

Low-lying states associated with the $i_{11/2}$ and $j_{15/2}$ orbitals are known from nuclear-reaction studies, but they are not populated in the beta decay.

The 2.13-MeV state that receives the primary beta decay is not so simply interpreted as the other states. It is to be associated with the promotion of a neutron from the $3p_{1/2}$ orbital below the 126 shell closure. The density (number per MeV) of states increases rapidly above this excitation, and the interpretations become more complex and less certain.

By suitable refinements, the spherical-shell model can be extended further from the doubly magic region. Primarily, it is necessary to drop the approximation that nucleons move independently in orbitals and to invoke a residual force, mainly short-range and attractive, between the nucleons. The spherical-shell model augmented by residual interactions can explain and correlate around the magic regions a large amount of data on binding energies, spins, magnetic moments, and the spectra of excited states.

The Collective Model

For nuclei more removed from the doubly magic regions, the spherical-shell model encounters difficulty in explaining the large observed electric quadrupole moments indicating cigar-shaped nuclei. For these nuclei a hybrid of liquid-drop and shell models, the collective model, has been proposed.

Nucleons can interact with one another in a collective fashion to deform the nuclear shape to a cigar shape. Such large spheroidal distortions are usual for nuclei far from magic, notably with $150 \lesssim A \lesssim 190$, and $224 \lesssim A$ (the symbol < denotes less than, and ~ means that the number is approximate). In these deformed regions the collective model prescribes that orbitals be computed in a cigar-shaped potential and that the relatively low-energy rotational excitations of the tumbling motion of the cigar shape be taken into account. The collective model has been highly successful in correlating and predicting nuclear properties in deformed regions. An example of a nuclear rotational band (a series of adjacent states) is provided by the decay of the isomer hafnium-180m, in Figure, through a cascade of gamma rays down the ground rotational band.

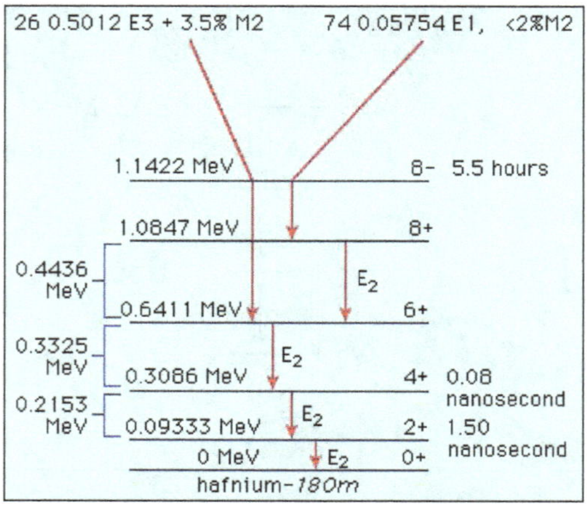

The decay scheme of hafnium-180m.

Radiochemical Analysis

Radiochemical Analysis is a branch of analytical, chemistry comprising an aggregate of methods for qualitatively determining the composition and content of radioisotopes in the products of transformations. Radioisotopes may arise from nuclear reactions both in natural substances and in specially irradiated materials. In contradistinction to radiometric assays, which determine the content of radioactive elements using physical instruments, radiochemical analysis establishes the content of radioisotopes in substances using chemical methods of separation and purification.

The identification and quantitative determination of radioisotopes are accomplished by measuring the gamma or alpha radioactivity of irradiated targets or of natural substances with alpha-ray or gamma-ray spectrometers. Radiometric apparatus permit the analysis of complex mixtures of radioisotopes without destruction of the original substance. In the analysis of substances containing a large number of radioisotopes or substances in which the relative concentrations of various radioisotopes vary greatly, the original substance is dissolved in water or acid; this is also done in cases where the decay of the radioisotope under study is accompanied by the emission of only beta particles or X rays. Isotopic or nonisotopic carriers are added to the solution and various chemical operations are carried out on the mixture in order to separate and purify the elements being investigated. The most common operations include methods of precipitation, extraction, chromatography, electrolysis, and distillation. The radioisotopes, which are isolated in radiochemically and chemically pure states, are then identified and their specific activities determined using radiometric spectrometers and counters of nuclear particles. The harmful effect of radioactive emissions requires adherence to special safety procedures.

Modern radiochemical analysis has been widely used in solving the many analytical problems that have arisen in connection with the production of nuclear fuel, the discovery and study of the properties of new radioactive elements and isotopes in activation analysis, and the investigations of the products of various nuclear reactions. Other uses of radiochemical analysis include the detection of the radioactive products of nuclear explosions on the surface of the earth and the study of the radioactivity of meteorites and surface layers of the moon induced by cosmic radiation.

Resonance Fluorescence

The phenomenon now known as the Mössbauer effect was first discovered by Rudolf Mössbauer in 1958 while he was a doctoral student at the Technical University of Munich. When he first published the result of his discovery, he referred to the effect that he had observed as 'Kernresonanzfluoreszenz von Gammastrahlung mit verschwindenden Rückstoβenergieverlustes'. Literally translated this means 'nuclear resonance fluorescence of gamma radiation with vanishing recoil energy loss' or (more succinctly) 'recoil-free nuclear resonance fluorescence'. This terminology uniquely defines the effect that Mössbauer sometimes described as 'like throwing a stone out of a rowboat in the wintertime'.

Atomic Resonance Fluorescence

The earliest observation of a phenomena that is based on atomic resonance fluorescence occurred in the early part of the 19th century. In 1802, the British scientist William Hyde Wollaston was observing the optical spectrum of the Sun and noticed the presence of dark lines in the continuum spectrum. A few years later, around 1814, Joseph von Fraunhofer observed the same phenomenon in the optical spectra of certain gases in the laboratory. A continuum source, such as a flame, showed narrow absorption lines when passed through a region of cold gas. In his investigations Fraunhofer identified the characteristic spectral lines associated with various elements.

In the following years, it was discovered that a gas heated to a high temperature would emit electromagnetic radiation at specific wavelengths giving rise to a spectrum consisting of bright lines at well-defined wavelengths on a dark background. In 1859 Gustav

Kirchhoff proposed what is now known as Kirchhoff's law (or sometimes Kirchhoff's law of spectroscopy, to distinguish it from the well-known Kirchhoff's laws related to currents in electrical circuits). Kirchhoff's law states that an element which, as a cold gas, produces an absorption spectrum with dark lines at particular wavelengths will produce an emission spectrum when heated with emission lines at those same wavelengths. This behavior is illustrated in figure where the properties of the resulting spectrum can be related to the details of the light source. Nearly a century after Wollaston's observation of solar spectral absorption lines, the relationship between the discrete spectra associated with specific elements was understood on the basis of the electronic structure of atoms.

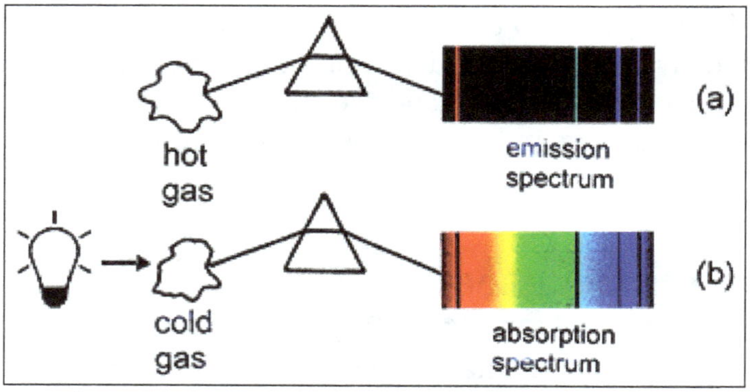

Observation of (a) An emission spectrum from a hot gas, and (b) An absorption spectrum from a cold gas.

Figure shows a slightly different arrangement for obtaining optical spectra that provides considerable insight into the nature of atomic transitions. At point (a) in figure an absorption spectrum is observed as in figure (b). However, an emission spectrum is observed at point (b) in figure. These two different spectra can be explained in the following way. Photons from the continuum source, which have just the right energy, will excite atomic transitions in the gas atoms. In this way, the energy of the photon will be absorbed, thus preferentially eliminating that particular wavelength from the continuum spectrum, giving rise to the dark lines in the spectrum observed in the forward direction from the source to the spectrum at point (a). The atoms that have excited electronic levels as a result of absorbing photons are unstable and will spontaneously decay back to their ground state. In this process, the atom will release a photon of the same energy (more or less) as the photon that had been absorbed. The question of how closely the absorbed and emitted photon energies are to one another should be considered. The absorbed photons were traveling in a line from the source to point (a), however, the re-emitted photons are re-emitted isotopically, that is, uniformly in all spatial directions. These re-emitted photons give rise to the emission spectrum at point (b). They also contribute to the wavelengths of the absorption lines at point (a). However, as only a small fraction of the re-emitted photons will be traveling along the line from the source to point (a), then there is a net decrease caused by the absorption at these wavelengths in this direction, giving rise to the dark lines.

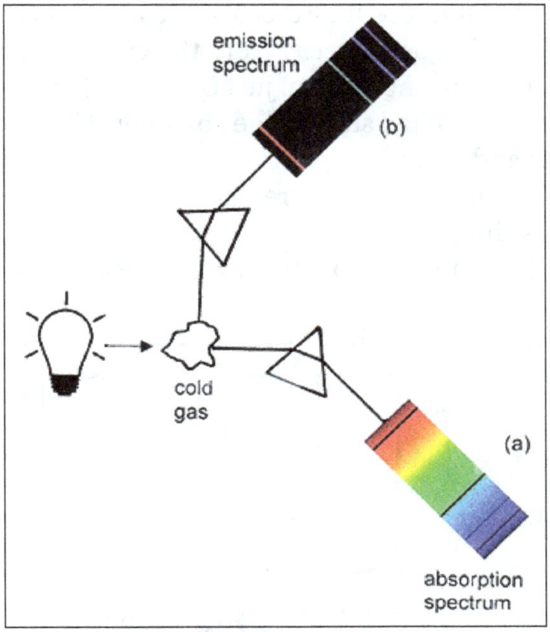

Geometry illustrating (a) Absorption spectrum of the continuum spectrum passing through cold gas, and (b) The spectrum of re-emitted photons, as described in the text.

In 1904, Robert W Wood elaborated on the concept of the emission spectrum shown at point (b) in figure and described the phenomenon of atomic resonance fluorescence, as illustrated in figure. In figure (a) a hot gas (the source) produces an emission spectrum characteristic of the element in the gas that is then observed as bright emission lines. In figure (b) a sample of cold gas of the same element (the absorber) is placed between the hot gas and the spectrometer. Since the atomic energy levels in the absorber are the same as they are in the source, the cold gas will preferentially absorb the photons produced by the source, thus eliminating (more or less) the bright emission lines in the resulting spectrum.

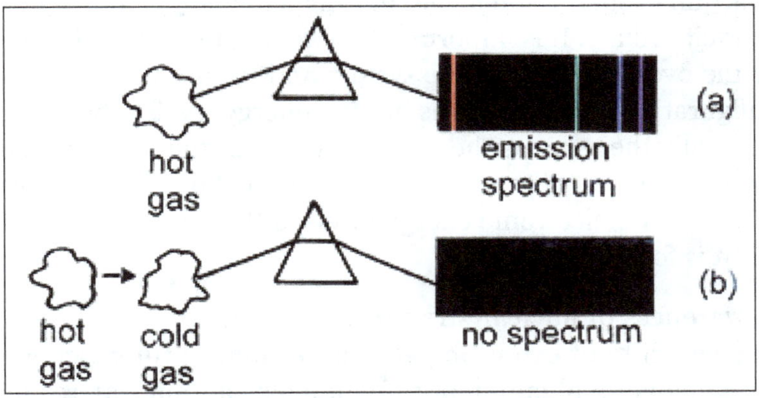

An atomic resonance fluorescence experiment. (a) Emission spectrum produced by the source of hot gas is observed by the spectrometer, and (b) Emission spectrum from the source is absorbed by the absorber and greatly reduces the spectrum observed by the spectrometer.

Figure shows the atomic resonance fluorescence phenomenon from an atomic standpoint. The source atom undergoes a decay from the excited state to the ground state, thereby emitting a photon. The photon is of just the correct energy to excite a transition from the ground state to the excited state in the absorber. This excited state then decays back to the ground state and re-emits a photon of the same energy. The re-emitted photon is given off in a random (isotropic) direction, thus decreasing the flux of photons in the direction towards the spectrometer. This phenomenon is fairly straightforward to observe and is a clear demonstration of the quantized nature of the atomic energy levels.

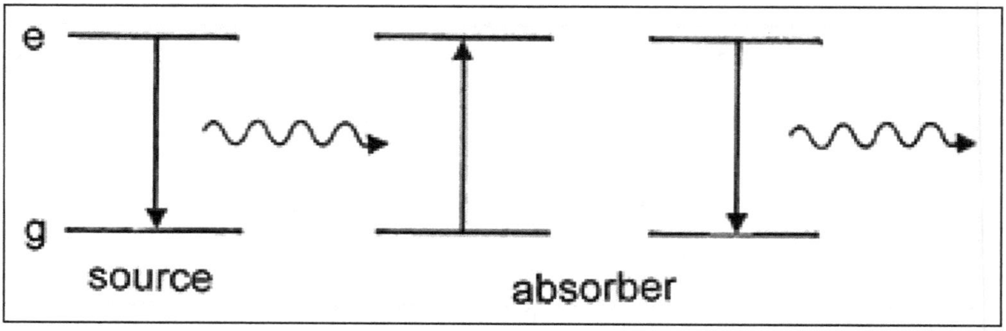

Atomic energy levels and transitions in the source and absorber during atomic resonance fluorescence; g = ground state and e = excited state.

Nuclei, like atoms, also have well-defined energy levels. This is because a nucleus is a bound system of particles (i.e. the neutrons and protons, collectively referred to as nucleons) just as the atom is a bound system of the nucleus and its electrons. Because both the neutron and the proton are fermions, they must occupy the energy levels in the nucleus in a way that satisfies the Pauli exclusion principle, just as the electrons in the atom must obey this principle. Thus, the lowest energy configuration of neutrons and protons in their energy levels that satisfies the Pauli principle represents the ground state of the nucleus, analogous to the situation for electrons in the atom. If a neutron and/or proton acquires energy and moves to a higher energy level, then the nucleus can be raised to an excited state. It is important to note that the energy levels of the nucleus are defined by the overall energy of the system and these energies are determined by the overall configuration of the nucleons in their energy levels. The energy levels of the nucleus are not really the same as the energy levels that the neutrons and protons can occupy but are the net result of the total configuration of particles. The same is true for atomic energy levels and electronic energy levels, although here the distinction is not as important as it is for nuclei.

The growing awareness of nuclear structure during the middle of the 20th century suggested the possibility of extending atomic resonance fluorescence experiments to consider an analogous phenomenon for nuclei, i.e. nuclear resonance fluorescence.

References

- What-are-radioactive-materials: nibusinessinfo.co.uk, Retrieved 8 August, 2019
- What-radioactive-waste: nibusinessinfo.co.uk, Retrieved 9 April, 2019
- Radiochemistry: thefreedictionary.com, Retrieved 11 January, 2019
- Radiochemical-methods: pace.edu, Retrieved 13 May, 2019
- Radioactive-isotope, science: britannica.com, Retrieved 25 February, 2019
- Radioactivity, science: britannica.com, Retrieved 5 June, 2019
- Radiochemical-Analysis: thefreedictionary.com, Retrieved 3 March, 2019

4

Nuclear Reactions

The process in which two nuclear particles interact and produce two or more nuclear particles is called a nuclear reaction. Some of the areas of study related to nuclear reactions are photonuclear reaction, energetics of nuclear reactions and compound nucleus reactions. These aspects of nuclear reactions have been thoroughly discussed in this chapter.

A nuclear reaction is considered to be the process in which two nuclear particles (two nuclei or a nucleus and a nucleon) interact to produce two or more nuclear particles or γ-rays (gamma rays). Thus, a nuclear reaction must cause a transformation of at least one nuclide to another. Sometimes if a nucleus interacts with another nucleus or particle without changing the nature of any nuclide, the process is referred to a nuclear scattering, rather than a nuclear reaction. Perhaps the most notable nuclear reactions are the nuclear fusion reactions of light elements that power the energy production of stars and the Sun. Natural nuclear reactions occur also in the interaction between cosmic rays and matter.

The most notable man-controlled nuclear reaction is the fission reaction which occurs in nuclear reactors. Nuclear reactors are devices to initiate and control a nuclear chain reaction, but there are not only manmade devices. The world's first nuclear reactor operated about two billion years ago. The natural nuclear reactor formed at Oklo in Gabon, Africa, when a uranium-rich mineral deposit became flooded with groundwater that acted as a neutron moderator, and a nuclear chain reaction started. These fission reactions were sustained for hundreds of thousands of years, until a chain reaction could no longer be supported. This was confirmed by existence of isotopes of the fission-product gas xenon and by different ratio of U-235/U-238 (enrichment of natural uranium).

Notation of Nuclear Reactions

Standard nuclear notation shows the chemical symbol, the mass number and the atomic number of the isotope.

If the initial nuclei are denoted by *a* and *b*, and the product nuclei are denoted by *c* and *d*, the reaction can be represented by the equation:

$$a + b \rightarrow c + d$$

$$^{10}_{5}B + ^{1}_{0}n \rightarrow ^{7}_{3}Li + ^{4}_{2}He + 2.8Mev$$

This equation describes neutron capture in the boron, which is diluted in the coolant. Boric acid is used in nuclear power plants as a long-term compensator of nuclear fuel reactivity.

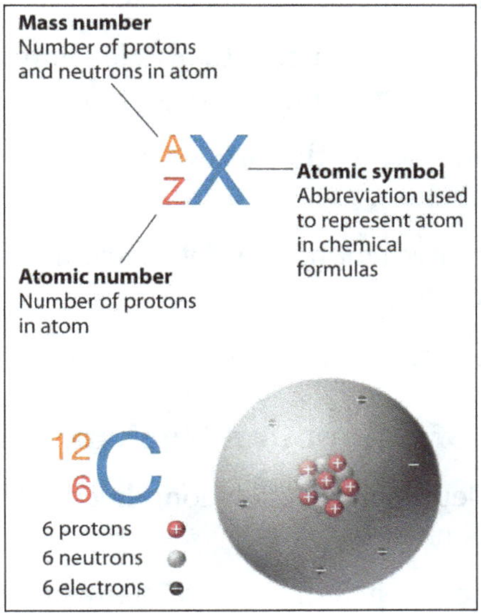

Instead of using the full equations in the style above, in many situations a compact notation is used to describe nuclear reactions. This style of the form a(b,c)d is equivalent to a + b producing c + d. Light particles are often abbreviated in this shorthand, typically p means proton, n means neutron, d means deuteron, α means an alpha particle or helium-4, β means beta particle or electron, γ means gamma photon, etc. The reaction above would be written as 10B(n,α)7Li.

Basic Classification of Nuclear Reactions

In order to understand the nature of neutron nuclear reactions, the classification according to the time scale of of these reactions has to be introduced. Interaction time is critical for defining the reaction mechanism.

There are two extreme scenarios for nuclear reactions (not only neutron reactions):

- A projectile and a target nucleus are within the range of nuclear forces for the very short time allowing for an interaction of a single nucleon only. These type of reactions are called the direct reactions.
- A projectile and a target nucleus are within the range of nuclear forces for the time allowing for a large number of interactions between nucleons. These type of reactions are called the compound nucleus reactions.

In fact, there is always some non-direct (multiple internuclear interaction) component in all reactions, but the direct reactions have this component limited.

Basic Characteristics of Direct Reactions

- The direct reactions are fast and involve a single-nucleon interaction.
- The interaction time must be very short ($\sim 10^{-22}$ s).
- The direct reactions require incident particle energy larger than ~ 5 MeV/Ap. (Ap is the atomic mass number of a projectile.)
- Incident particles interact on the surface of a target nucleus rather than in the volume of a target nucleus.
- Products of the direct reactions are not distributed isotropically in angle, but they are forward focused.
- Direct reactions are of importance in measurements of nuclear structure.

Basic Characteristics of Compound Nucleus Reactions

- The compound nucleus is a relatively long-lived intermediate state of particle-target composite system.
- The compound nucleus reactions involve many nucleon-nucleon interactions.
- The large number of collisions between the nucleons leads to a thermal equilibrium inside the compound nucleus.
- The time scale of compound nucleus reactions is of the order of 10^{-18} s – 10^{-16} s.
- The compound nucleus reactions is usually created if the projectile has low energy.
- Incident particles interact in the volume of a target nucleus.
- Products of the compound nucleus reactions are distributed near isotropically in angle (the nucleus loses memory of how it was created – the Bohr's hypothesis of independence).
- The mode of decay of compound nucleus do not depend on the way the compound nucleus is formed.
- Resonances in the cross-section are typical for the compound nucleus reaction.

Types of Nuclear Reactions

Although the number of possible nuclear reactions is enormous, nuclear reactions can be sorted by types. Most of nuclear reactions are accompanied by gamma emission. Some examples are:

- Elastic scattering: Occurs, when no energy is transferred between the target nucleus and the incident particle.

  208Pb (n, n) 208Pb

- Inelastic scattering: Occurs, when energy is transferred. The difference of kinetic energies is saved in excited nuclide.

  40Ca (α, α') 40mCa

- Capture reactions: Both charged and neutral particles can be captured by nuclei. This is accompanied by the emission of γ -rays. Neutron capture reaction produces radioactive nuclides (induced radioactivity).

  238U (n, γ) 239U

- Transfer Reactions: The absorption of a particle accompanied by the emission of one or more particles is called the transfer reaction.

  4He (α, p) 7Li

- Fission reactions: Nuclear fission is a nuclear reaction in which the nucleus of an atom splits into smaller parts (lighter nuclei). The fission process often produces free neutrons and photons (in the form of gamma rays), and releases a large amount of energy.

  235U (n, 3 n) fission products

- Fusion reactions: Occur when, two or more atomic nuclei collide at a very high speed and join to form a new type of atomic nucleus. The fusion reaction of deuterium and tritium is particularly interesting because of its potential of providing energy for the future.

  3T (d, n) 4He

- Spallation reactions: Occur, when a nucleus is hit by a particle with sufficient energy and momentum to knock out several small fragments or, smash it into many fragments.

- Nuclear decay (Radioactive decay): Occurs when an unstable atom loses energy by emitting ionizing radiation. Radioactive decay is a random process at the level of single atoms, in that, according to quantum theory, it is impossible to predict when a particular atom will decay. There are many types of radioactive decay:

  - Alpha radioactivity: Alpha particles consist of two protons and two neutrons bound together into a particle identical to a helium nucleus. Because of its very large mass (more than 7000 times the mass of the beta particle) and its charge, it heavy ionizes material and has a very short range.

  - Beta radioactivity: Beta particles are high-energy, high-speed electrons or positrons emitted by certain types of radioactive nuclei such as potassium-40.

The beta particles have greater range of penetration than alpha particles, but still much less than gamma rays. The beta particles emitted are a form of ionizing radiation also known as beta rays. The production of beta particles is termed beta decay.

- Gamma radioactivity: Gamma rays are electromagnetic radiation of an very high frequency and are therefore high energy photons. They are produced by the decay of nuclei as they transition from a high energy state to a lower state known as gamma decay. Most of nuclear reactions are accompanied by gamma emission.
- Neutron emission: Neutron emission is a type of radioactive decay of nuclei containing excess neutrons (especially fission products), in which a neutron is simply ejected from the nucleus. This type of radiation plays key role in nuclear reactor control, because these neutrons are delayed neutrons.

Decay Type	Radiation Emitted	Generic Equation	Model
Alpha decay	$^{4}_{2}\alpha$	$^{A}_{Z}X \longrightarrow {}^{A-4}_{Z-2}X' + {}^{4}_{2}\alpha$	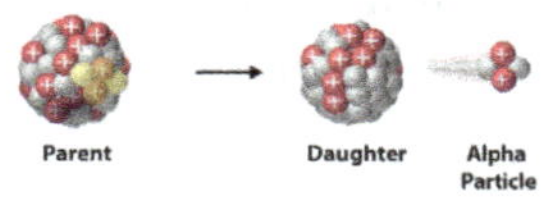
Beta decay	$^{0}_{-1}\beta$	$^{A}_{Z}X \longrightarrow {}^{A}_{Z+1}X' + {}^{0}_{-1}\beta$	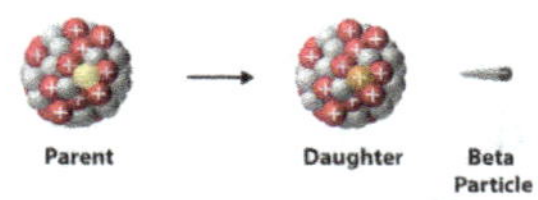
Positron emission	$^{0}_{+1}\beta$	$^{A}_{Z}X \longrightarrow {}^{A}_{Z-1}X' + {}^{0}_{+1}\beta$	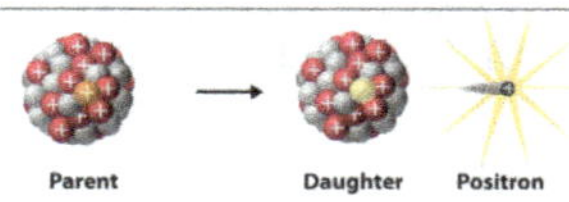
Electron capture	X rays	$^{A}_{Z}X + {}^{0}_{-1}e \longrightarrow {}^{A}_{Z-1}X' + \text{X ray}$	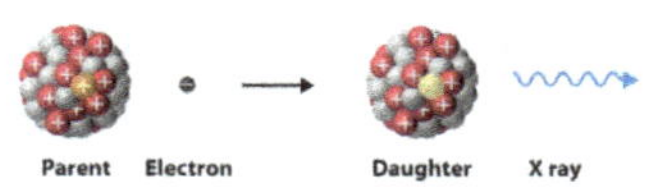
Gamma emission	$^{0}_{0}\gamma$	$^{A}_{Z}X^{*} \xrightarrow{\text{Relaxation}} {}^{A}_{Z}X' + {}^{0}_{0}\gamma$	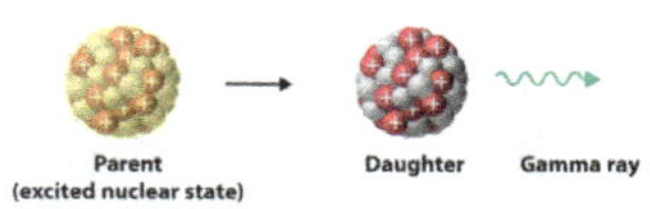
Spontaneous fission	Neutrons	$^{A+B+C}_{Z+Y}X \longrightarrow {}^{A}_{Z}X' + {}^{B}_{Y}X' + C{}^{1}_{0}n$	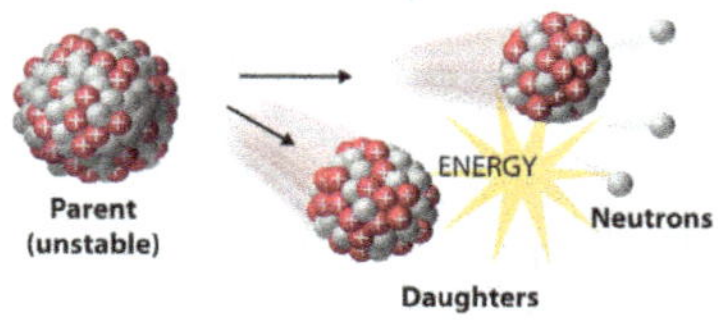

Radioactive decays.

Energetics of Nuclear Reactions – Q-value

In nuclear and particle physics the energetics of nuclear reactions is determined by the Q-value of that reaction. The Q-value of the reaction is defined as the difference between the sum of the masses of the initial reactants and the sum of the masses of the final products, in energy units (usually in MeV).

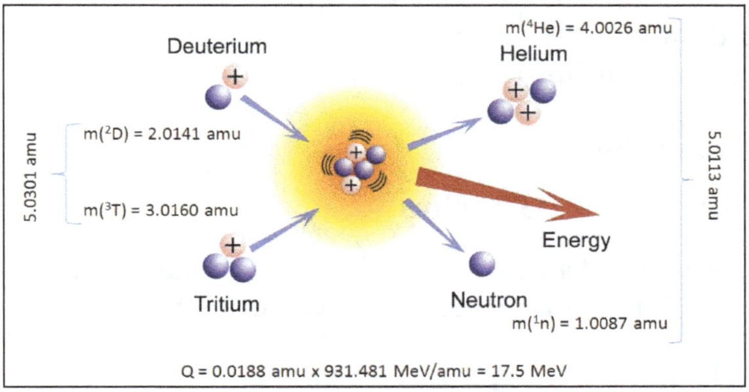

Q-value of DT fusion reaction.

Consider a typical reaction, in which the projectile a and the target A gives place to two products, B and b. This can also be expressed in the notation that we used so far, a + A → B + b, or even in a more compact notation, A(a,b)B.

The Q-value of this reaction is given by:

$$Q = [m_a + m_A - (m_b + m_B)]c^2$$

Which is the same as the excess kinetic energy of the final products:

$$Q = T_{final} - T_{initial}$$

$$= T_b + T_B - (T_a + T_A)$$

For reactions in which there is an increase in the kinetic energy of the products Q is positive. The positive Q reactions are said to be exothermic (or exergic). There is a net release of energy, since the kinetic energy of the final state is greater than the kinetic energy of the initial state.

For reactions in which there is a decrease in the kinetic energy of the products Q is negative. The negative Q reactions are said to be endothermic (or endoergic) and they require a net energy input.

The energy released in a nuclear reaction can appear mainly in one of three ways:

- Kinetic energy of the products.
- Emission of gamma rays. Gamma rays are emitted by unstable nuclei in their transition from a high energy state to a lower state known as gamma decay.
- Metastable state. Some energy may remain in the nucleus, as a metastable energy level.

A small amount of energy may also emerge in the form of X-rays. Generally, products of nuclear reactions may have different atomic numbers, and thus the configuration of their electron shells is different in comparison with reactants. As the electrons rearrange themselves and drop to lower energy levels, internal transition X-rays (X-rays with precisely defined emission lines) may be emitted.

Exothermic Reactions

Exothermic Reaction - DT fusion

The DT fusion reaction of deuterium and tritium is particularly interesting because of its potential of providing energy for the future. Calculate the reaction Q-value.

3T (d, n) 4He

The atom masses of the reactants and products are:

$m(^3T) = 3.0160$ amu

$m(^2D) = 2.0141$ amu

$m(^1n) = 1.0087$ amu

$m(^4He) = 4.0026$ amu

Using the mass-energy equivalence, we get the Q-value of this reaction as:

$$Q = \{(3.0160+2.0141)\ [amu] - (1.0087+4.0026)\ [amu]\} \times 931.481\ [MeV/amu]$$

$$= 0.0188 \times 931.481 = 17.5\ MeV$$

Exothermic Reaction - Tritium in Reactors

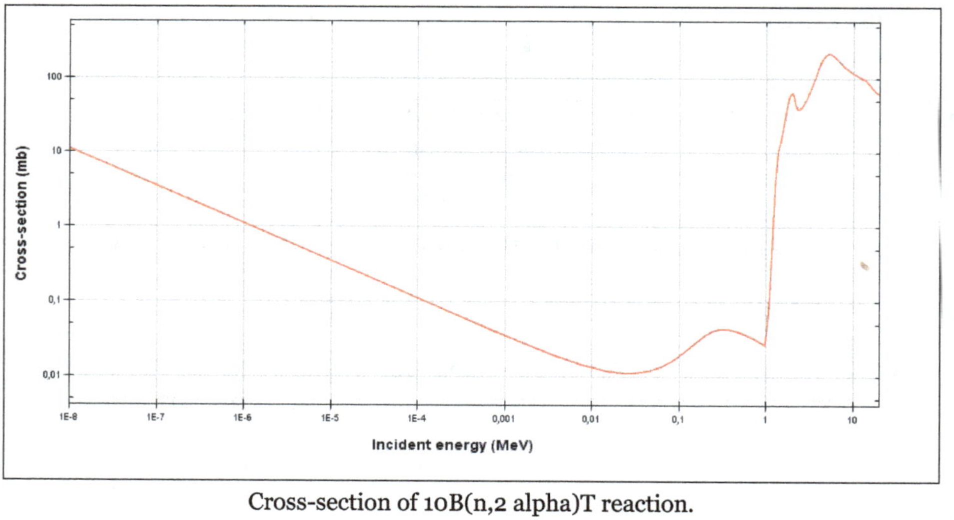

Cross-section of 10B(n,2 alpha)T reaction.

Tritium is a byproduct in nuclear reactors. Most of the tritium produced in nuclear power plants stems from the boric acid, which is commonly used as a chemical shim to compensate an excess of initial reactivity. Main reaction, in which the tritium is generated from boron is below:

10B(n,2*alpha)T

This reaction of a neutron with an isotope ^{10}B is the main way, how radioactive tritium in primary circuit of all PWRs is generated. Note that, this reaction is a threshold reaction due to its cross-section.

Calculate the reaction Q-value.

The atom masses of the reactants and products are:

$m(^{10}B)$ = 10.01294 amu

$m(^{1}n)$ = 1.00866 amu

$m(^{3}T)$ = 3.01604 amu

$m(^{4}He)$ = 4.0026 amu

Using the mass-energy equivalence, we get the Q-value of this reaction as:

Q = {(10.0129+1.00866) [amu] – (3.01604+2 x 4.0026) [amu]} x 931.481 [MeV/amu]

= 0.00036 x 931.481 = 0.335 MeV

Endothermic Reactions

Endothermic Reaction - Photoneutrons

In nuclear reactors the gamma radiation plays a significant role also in reactor kinetics and in a subcriticality control. Especially in nuclear reactors with D_2O moderator (CANDU reactors) or with Be reflectors (some experimental reactors). Neutrons can be produced also in (γ, n) reactions and therefore they are usually referred to as photoneutrons.

A high energy photon (gamma ray) can under certain conditions eject a neutron from a nucleus. It occurs when its energy exceeds the binding energy of the neutron in the nucleus. Most nuclei have binding energies in excess of 6 MeV, which is above the energy of most gamma rays from fission. On the other hand there are few nuclei with sufficiently low binding energy to be of practical interest. These are: ^{2}D, ^{9}Be, ^{6}Li, ^{7}Li and ^{13}C. As can be seen from the table the lowest threshold have ^{9}Be with 1.666 MeV and ^{2}D with 2.226 MeV.

Table: Nuclides with low photodisintegration threshold energies.

Nuclide	Threshold(MeV)	Reaction
^{2}D	2.225	$^{2}H(\gamma,n)^{1}H$
^{6}Li	3.697	$^{6}Li(\gamma,n+p)^{4}He$
^{6}Li	5.67	$^{6}Li(\gamma,n)^{5}Li$
^{7}Li	7.251	$^{7}Li(\gamma,n)^{6}Li$
^{9}Be	1.667	$^{9}Be(\gamma,n)^{8}Be$
^{13}C	4.9	$^{13}C(\gamma,n)^{12}C$

In case of deuterium, neutrons can be produced by the interaction of gamma rays (with a minimum energy of 2.22 MeV) with deuterium:

$$^{2}_{1}D+\gamma\rightarrow^{1}_{1}H+n$$

The reaction Q-value is calculated below.

The atom masses of the reactant and products are:

m(^{2}D) = 2.01363 amu

m(^{1}n) = 1.00866 amu

m(^{1}H) = 1.00728 amu

Using the mass-energy equivalence, we get the Q-value of this reaction as:

Q = {2.01363 [amu] – (1.00866+1.00728) [amu]} x 931.481 [MeV/amu]

= -0.00231 x 931.481 = -2.15 MeV

Endothermic Reaction - (α,n) Reaction

Calculate the reaction Q-value of the following reaction:

7Li (α, n) 10B

The atom masses of the reactants and products are given below.

m(^{4}He) = 4.0026 amu

m(^{7}Li) = 7.0160 amu

m(^{1}n) = 1.0087 amu

m(^{10}B) = 10.01294 amu

Using the mass-energy equivalence, we get the Q-value of this reaction as:

Q = {(7.0160+4.0026) [amu] – (1.0087+10.01294) [amu]} x 931.481 [MeV/amu]

= 0.00304 x 931.481 = -2.83 MeV

Types of Particles in Nuclear Reactions

Many entities can be involved in nuclear reactions. The most common are protons, neutrons, alpha particles, beta particles, positrons, and gamma rays, as shown in table. Protons ($^1_1$p, also represented by the symbol $^1_1$H) and neutrons (^1_0n) are the constituents of atomic nuclei. Alpha particles ($^4_2$He, also represented by the symbol ^4_2a) are high-energy helium nuclei. Beta particles ($^0_{-1}\beta$, also represented by the symbol $^0_{-1}$e) are high-energy electrons, and gamma rays are photons of very high-energy electromagnetic radiation. Positrons ($^0_{+1}$e, also represented by the symbol $^0_{+1}\beta$) are positively charged electrons ("anti-electrons"). The subscripts and superscripts are necessary for balancing nuclear equations, but are usually optional in other circumstances. For example, an alpha particle is a helium nucleus (He) with a charge of +2 and a mass number of 4, so it is symbolized $^4_2$He. This works because, in general, the ion charge is not important in the balancing of nuclear equations.

Table: Although many species are encountered in nuclear reactions, this table summarizes the names, symbols, representations, and descriptions of the most common of these.

Name	Symbol(s)	Representation	Description
Alpha particle	$^4_2$He or ^4_2a		(High-energy) helium nuclei consisting of two protons and two neutrons.
Beta particle	$^0_1$e or $^0_{-1}\beta$		(High-energy) electrons.
Positron	$^0_{+1}$e or $^0_{+1}\beta$		Particles with the same mass as an electron but with 1 unit of positive charge.
Proton	$^1_1$H or $^1_1$p		Nuclei of hydrogen atoms.
Neutron	$^1_0$n		Particles with a mass approximately equal to that of a proton but with no charge.
Gamma ray	γ		Very high-energy electromagnetic radiation.

Note that positrons are exactly like electrons, except they have the opposite charge. They are the most common example of antimatter, particles with the same mass but the opposite state of another property (for example, charge) than ordinary matter. When antimatter encounters ordinary matter, both are annihilated and their mass is converted into energy in the form of gamma rays (γ)—and other much smaller subnuclear particles according to the mass-energy equivalence equation $E = mc^2$, seen in the preceding section. For example, when a positron and an electron collide, both are annihilated and two gamma ray photons are created:

$$^{0}_{-1}\text{e} + ^{0}_{+1}\text{e} \rightarrow \gamma + \gamma$$

Gamma rays compose short wavelength, high-energy electromagnetic radiation and are (much) more energetic than better-known X-rays that can behave as particles in the wave-particle duality sense. Gamma rays are produced when a nucleus undergoes a transition from a higher to a lower energy state, similar to how a photon is produced by an electronic transition from a higher to a lower energy level. Due to the much larger energy differences between nuclear energy shells, gamma rays emanating from a nucleus have energies that are typically millions of times larger than electromagnetic radiation emanating from electronic transitions.

Balancing Nuclear Reactions

A balanced chemical reaction equation reflects the fact that during a chemical reaction, bonds break and form, and atoms are rearranged, but the total numbers of atoms of each element are conserved and do not change. A balanced nuclear reaction equation indicates that there is a rearrangement during a nuclear reaction, but of subatomic particles rather than atoms. Nuclear reactions also follow conservation laws, and they are balanced in two ways:

- The sum of the mass numbers of the reactants equals the sum of the mass numbers of the products.
- The sum of the charges of the reactants equals the sum of the charges of the products.

If the atomic number and the mass number of all but one of the particles in a nuclear reaction are known, we can identify the particle by balancing the reaction. For instance, we could determine that $^{17}_{8}\text{O}$ is a product of the nuclear reaction of $^{14}_{7}\text{N}$ and $^{4}_{2}\text{He}$ if we knew that a proton, $^{1}_{1}\text{H}$, was one of the two products.

Following are the equations of several nuclear reactions that have important roles in the history of nuclear chemistry:

- The first naturally occurring unstable element that was isolated, polonium, was discovered by the Polish scientist Marie Curie and her husband Pierre in 1898. It decays, emitting α particles:

  $$^{212}_{84}\text{Po} \rightarrow ^{208}_{82}\text{Pb} + ^{4}_{2}\text{He}$$

- The first nuclide to be prepared by artificial means was an isotope of oxygen,

^{17}O. It was made by Ernest Rutherford in 1919 by bombarding nitrogen atoms with α particles:

$$^{14}_{7}N + ^{4}_{2}a \rightarrow ^{17}_{8}O + ^{1}_{1}H$$

- James Chadwick discovered the neutron in 1932, as a previously unknown neutral particle produced along with ^{12}C by the nuclear reaction between ^{9}Be and ^{4}He:

$$^{9}_{4}Be + ^{4}_{2}He \rightarrow ^{12}_{6}C + ^{1}_{0}n$$

- The first element to be prepared that does not occur naturally on the earth, technetium, was created by bombardment of molybdenum by deuterons (heavy hydrogen, $^{2}_{1}H$), by Emilio Segre and Carlo Perrier in 1937:

$$^{2}_{1}H + ^{97}_{42}Mo \rightarrow 2^{1}_{0}n + ^{97}_{43}Tc$$

- The first controlled nuclear chain reaction was carried out in a reactor at the University of Chicago in 1942. One of the many reactions involved was:

$$^{235}_{92}U + ^{1}_{0}n \rightarrow ^{87}_{35}Br + ^{146}_{57}La + 3^{1}_{0}n$$

Energetics of Nuclear Reactions

Nuclear reactions, like chemical reactions, are accompanied by changes in energy. The energy changes in nuclear reactions, however, are enormous compared with those of even the most energetic chemical reactions. In fact, the energy changes in a typical nuclear reaction are so large that they result in a measurable change of mass.

Mass–Energy Balance

The relationship between mass (m) and energy (E) is expressed in the following equation:

$$E = mc^2$$

Where,

- c is the speed of light (2.998×10^8 m/s),
- E and m are expressed in units of joules and kilograms, respectively.

Albert Einstein first derived this relationship in 1905 as part of his special theory of relativity: the mass of a particle is directly proportional to its energy. Thus according to Equation $E = mc^2$, every mass has an associated energy, and similarly, any reaction that involves a change in energy must be accompanied by a change in mass. This implies that all exothermic reactions should be accompanied by a decrease in mass, and all endothermic reactions should be accompanied by an increase in mass. Given the law of conservation of mass, how can this be true? The solution to this apparent contradiction is that chemical reactions are indeed accompanied by changes in mass, but these changes are simply too small to be detected. As you may recall, all particles exhibit wavelike behavior, but the wavelength is inversely proportional to the mass of the particle (actually, to its momentum, the product of its mass and velocity). Consequently, wavelike behavior is detectable only for particles with very small masses, such as electrons. For example, the chemical equation for the combustion of graphite to produce carbon dioxide is as follows:

$$\mathrm{C(graphite)} + \frac{1}{2}\mathrm{O_2(g)} \rightarrow \mathrm{CO_2(g)}\ \Delta\mathrm{H}^\circ = -393.5\,\mathrm{kJ/mol}$$

Combustion reactions are typically carried out at constant pressure, and under these conditions, the heat released or absorbed is equal to ΔH. When a reaction is carried out at constant volume, the heat released or absorbed is equal to ΔE. For most chemical reactions, however, $\Delta E \approx \Delta H$. If we rewrite Einstein's equation as,

$$\Delta \mathrm{E} = (\Delta \mathrm{m})\mathrm{c}^2$$

we can rearrange the equation to obtain the following relationship between the change in mass and the change in energy:

$$\Delta m = \frac{\Delta E}{c^2}$$

Because 1 J = 1 (kg · m²)/s², the change in mass is as follows:

$$\Delta m \frac{-393.5kJ/\mathrm{mol}}{(2.998\times10^8 m/s)^2} = \frac{-3.935\times10^5(kg\cdot\mathrm{m}^2)/(\mathrm{s}^2\cdot\mathrm{mol})}{(2.998\times10^8\ \mathrm{m/s})^2} = -4.38\times10^{-12}\ \mathrm{kg/mol}$$

This is a mass change of about 3.6×10^{-10} g/g carbon that is burned, or about 100-millionths of the mass of an electron per atom of carbon. In practice, this mass change is much too small to be measured experimentally and is negligible.

In contrast, for a typical nuclear reaction, such as the radioactive decay of ^{14}C to ^{14}N and an electron (a β particle), there is a much larger change in mass:

$$^{14}\mathrm{C} \rightarrow {}^{14}\mathrm{N} + {}^{0}_{-1}\beta$$

We can use the experimentally measured masses of subatomic particles and common isotopes to calculate the change in mass directly. The reaction involves the conversion of a neutral ^{14}C atom to a positively charged ^{14}N ion (with six, not seven, electrons) and a negatively charged β particle (an electron), so the mass of the products is identical to the mass of a neutral ^{14}N atom. The total change in mass during the reaction is therefore the difference between the mass of a neutral ^{14}N atom (14.003074 amu) and the mass of a ^{14}C atom (14.003242 amu):

$$\Delta m = \text{mass}_{\text{products}} - \text{mass}_{\text{reactants}}$$
$$= 14.003074\,\text{amu} - 14.003242\,\text{amu} = -0.000168\,\text{amu}$$

The difference in mass, which has been released as energy, corresponds to almost one-third of an electron. The change in mass for the decay of 1 mol of ^{14}C is −0.000168 g = -1.68×10^{-4} g = -1.68×10^{-7} kg. Although a mass change of this magnitude may seem small, it is about 1000 times larger than the mass change for the combustion of graphite. The energy change is as follows:

$$\Delta E = (\Delta m)c^2 = -(1.68 \times 10^{-7}\,\text{kg})(2.998 \times 10^8\,\text{m/s})^2$$
$$= -1.51 \times 10^{10} (\text{kg} \cdot \text{m}^2)/\text{s}^2 = -1.51 \times 10^{10} J = -1.51 \times 10^7\,\text{kJ}$$

The energy released in this nuclear reaction is more than 100,000 times greater than that of a typical chemical reaction, even though the decay of ^{14}C is a relatively low-energy nuclear reaction.

Because the energy changes in nuclear reactions are so large, they are often expressed in kiloelectronvolts (1 keV = 10^3 eV), megaelectronvolts (1 MeV = 10^6 eV), and even gigaelectronvolts (1 GeV = 10^9 eV) per atom or particle. The change in energy that accompanies a nuclear reaction can be calculated from the change in mass using the relationship 1 amu = 931 MeV. The energy released by the decay of one atom of ^{14}C is thus,

$$(-1.68 \times 10^{-4}\,\text{amu})\left(\frac{931\,\text{MeV}}{\text{amu}}\right) = -0.156\,\text{MeV} = -156\,\text{keV}$$

Nuclear Binding Energies

We have seen that energy changes in both chemical and nuclear reactions are accompanied by changes in mass. Einstein's equation, which allows us to interconvert mass and energy, has another interesting consequence: The mass of an atom is always less than the sum of the masses of its component particles. The only exception to this rule is hydrogen-1 (^1H), whose measured mass of 1.007825 amu is identical to the sum of the masses of a proton and an electron. In contrast, the experimentally measured mass of an atom of deuterium (^2H) is 2.014102 amu, although its calculated mass is 2.016490 amu:

$$m_{2\text{H}} = m_{\text{neutron}} + m_{\text{proton}} + m_{\text{electron}}$$
$$= 1.008665\,\text{amu} + 1.007276\,\text{amu} = 0.000549\,\text{amu} = 2.016490\,\text{amu}$$

The difference between the sum of the masses of the components and the measured atomic mass is called the mass defect of the nucleus. Just as a molecule is more stable than its isolated atoms, a nucleus is more stable (lower in energy) than its isolated components. Consequently, when isolated nucleons assemble into a stable nucleus, energy is released. According to Equation $\Delta m = \frac{\Delta E}{c^2}$, this release of energy must be accompanied by a decrease in the mass of the nucleus.

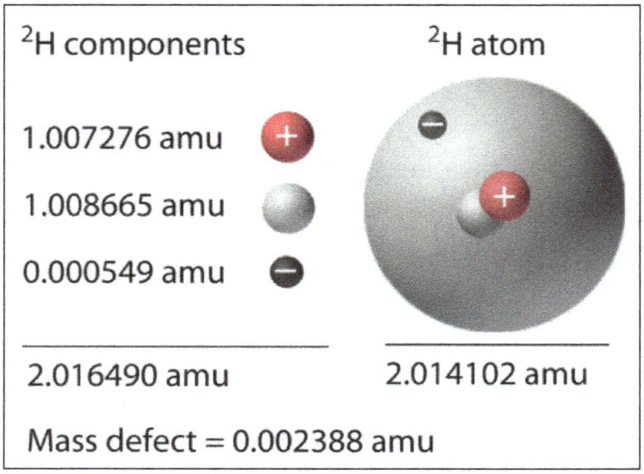

The figure above shows nuclear Binding Energy in Deuterium. The mass of a ^{2}H atom is less than the sum of the masses of a proton, a neutron, and an electron by 0.002388 amu; the difference in mass corresponds to the nuclear binding energy. The larger the value of the mass defect, the greater the nuclear binding energy and the more stable the nucleus.

The amount of energy released when a nucleus forms from its component nucleons is the nuclear binding energy. In the case of deuterium, the mass defect is 0.002388 amu, which corresponds to a nuclear binding energy of 2.22 MeV for the deuterium nucleus. Because the magnitude of the mass defect is proportional to the nuclear binding energy, both values indicate the stability of the nucleus.

Just as a molecule is more stable (lower in energy) than its isolated atoms, a nucleus is more stable than its isolated components.

Not all nuclei are equally stable. Chemists describe the relative stability of different nuclei by comparing the binding energy per nucleon, which is obtained by dividing the nuclear binding energy by the mass number (A) of the nucleus. As shown in Figure, the binding energy per nucleon increases rapidly with increasing atomic number until about Z = 26, where it levels off to about 8–9 MeV per nucleon and then decreases slowly. The initial increase in binding energy is not a smooth curve but exhibits sharp peaks corresponding to the light nuclei that have equal numbers of protons and neutrons (e.g., ^{4}He, ^{12}C, and ^{16}O). As mentioned earlier, these are particularly stable combinations.

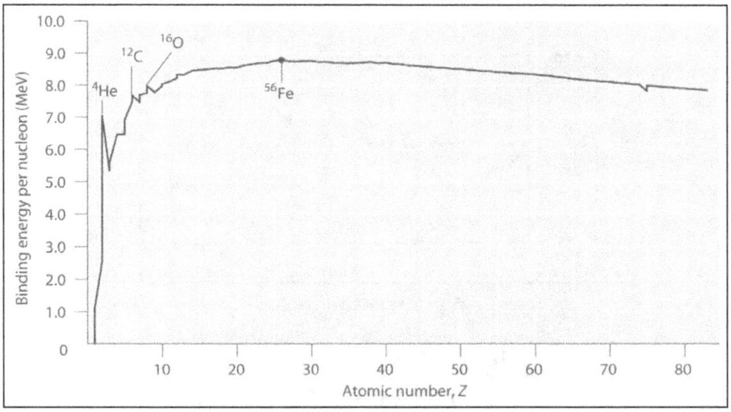

The Curve of Nuclear Binding Energy.

This plot of the average binding energy per nucleon as a function of atomic number shows that the binding energy per nucleon increases with increasing atomic number until about Z = 26, levels off, and then decreases. The sharp peaks correspond to light nuclei that have equal numbers of protons and neutrons.

Because the maximum binding energy per nucleon is reached at ^{56}Fe, all other nuclei are thermodynamically unstable with regard to the formation of ^{56}Fe. Consequently, heavier nuclei should spontaneously undergo reactions such as alpha decay, which result in a decrease in atomic number. Conversely, lighter elements should spontaneously undergo reactions that result in an increase in atomic number. This is indeed the observed pattern.

Photonuclear Reaction

Photonuclear Reaction, (also nuclear photoelectric effect) is the absorption of gamma-ray photons by atomic nuclei and the accompanying ejection of protons *p*, neutrons *n*, or heavier particles from the nuclei. The (γ, *p*) and (γ, *n*) photonuclear reactions have been studied the most; other reactions, such as (γ, *d*), (γ, *pn*), and (γ, *t*) are also known. For a proton or a neutron, which are called nucleons, to be ejected from an atomic nucleus, the gamma-ray photon energy E_γ must exceed the binding energy of the nucleon in the nucleus. The total cross section for all possible photonuclear reactions is called the gamma-ray absorption cross section for a nucleus.

For all except very light nuclei, the cross section σ_γ is small at high and low gamma-ray energies but has a high broad maximum, which is called a giant resonance, at some intermediate energy. As the mass number *A* of the nucleus increases, the position of the giant resonance decreases monotonically from 20–25 million electron volts (MeV) in light nuclei to 13 MeV in heavy nuclei.

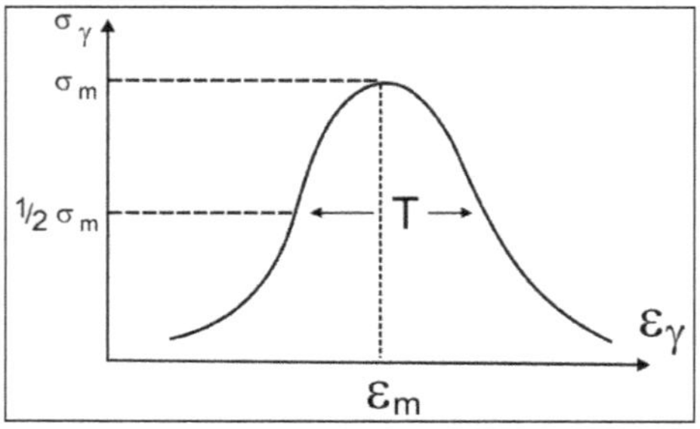

A giant resonance.

The relation between the energy E_m that corresponds to the resonance peak and A is described by the equation $E_m = 34A^{-1/6}$. The width Γ of the resonance is about 4–8 MeV; it is minimum for magic number nuclei—for example, $\Gamma(^{208}Pb) = 3.9$ MeV—and is maximum for deformed nuclei—for example, $\Gamma(^{165}Ho) = 7$ MeV. In the region of a giant resonance, the absorption curve is not monotonic but has a definite structure. For deformed nuclei, the absorption curve has a double peak. For light and medium nuclei, as well as for certain heavy nuclei, the absorption curve has several maxima with widths of hundreds of thousands of electron volts. The energy distribution of photoneutrons in a resonance region is nearly Maxwellian. At the same time, there are some deviations; for example, the number of neutrons in the high-energy region of the spectrum is larger than is expected for a Maxwellian distribution. In most cases, the distribution of photoprotons is not Maxwellian.

A giant resonance is associated with the excitation by gammaray photons of natural oscillations of protons relative to neutrons; such oscillations are known as dipole oscillations. Nucleons may escape directly from a nucleus during dipole oscillations but may be ejected only after the oscillations are damped. The ordered oscillations of nucleons gradually evolve into a somewhat complicated thermal motion. As a result, an excited compound nucleus is formed, from which protons or neutrons are "evaporated" The width Γ of a giant resonance is determined by the lifetime of the dipole oscillations. As a rule, when the gamma-ray energy exceeds the energy of a giant resonance, nucleons that absorb the gamma-ray photons escape rapidly from the nucleus. In this case, no dipole oscillations occur since the nucleus has not built them up, and the photonuclear reaction mechanism is direct. For example, at $\varepsilon_\gamma \geq 70$ MeV, the gamma-ray absorption mechanism becomes a two-nucleon mechanism. In addition to dipole oscillations in a nucleus, quadrupole, octupole, and other types of oscillations may be excited, but the role of such higher-order oscillations in photonuclear reactions is not substantial

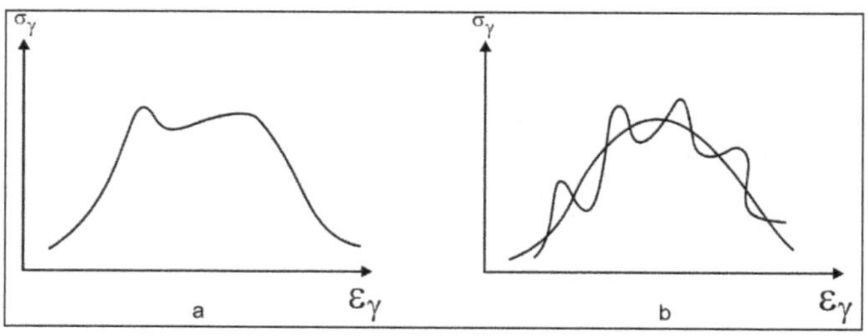

The fine structure of a giant resonance: (a) for deformed nuclei, (b) for spherical nuclei.

The term "photonuclear reactions" is sometimes applied to processes in which the absorption of high-energy gamma rays—that is, photons with energies of ~ 1.5×10^8 eV—by nuclei or individual nucleons results in the production of pions or other elementary particles. Examples of such reactions include $\gamma + p \rightarrow n + \pi^-$ and $\gamma + p \rightarrow p + \pi^o$.

Nuclear Fission and Fusion

Nuclear fission and nuclear fusion are two types of nuclear reactions. Fission means to split and fusion means to combine. Therefore in fission reaction a heavy nucleus of atom is bombarded so that it splits and in fusion reaction two lighter nuclei are heated at high temperature so that they combine to form a heavy nucleus.

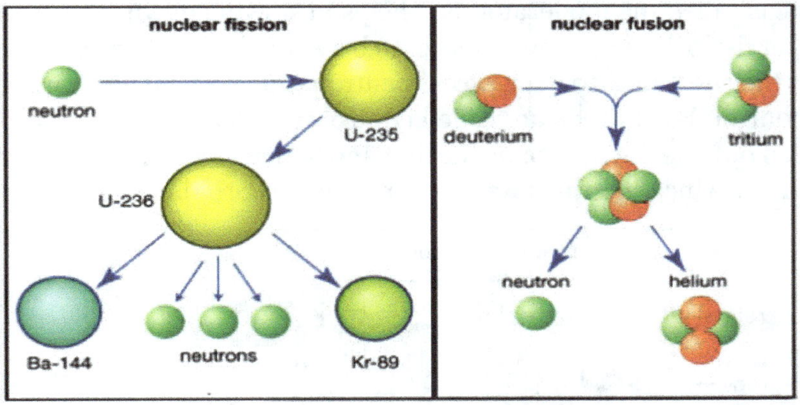

The mass of nucleus serves as the source of nuclear energy which is released mainly in the form of heat. There are two types of nuclear reaction. They are:

1. Nuclear Fission
2. Nuclear Fusion

Nuclear Fission

The heavy nucleus of radioactive atoms like uranium, plutonium or thorium is bombarded with low energy neutrons which split the nucleus into smaller nuclei. This process is called nuclear fission. For example when uranium-235 atoms are bombarded with neutrons then the heavy uranium nucleus splits to produce barium-139 and krypton-94 with the emission of three neutrons. A lot of energy is also produced in this reaction because mass is converted into energy.

$$\underset{\text{Heavy atom}}{{}^{235}_{92}U} + \underset{\text{Neutron}}{{}^{1}_{0}n} \xrightarrow{\text{Fission}} \underset{\text{Medium weight atom}}{{}^{139}_{56}Ba + {}^{94}_{36}Kr} + \underset{\text{Neutrons}}{3\,{}^{1}_{0}n} + \text{Tremendous amount of energy}$$

Also, in a nuclear fission reaction neutrons are used up and produced as well. The neutrons produced in the nuclear fission reaction leads to further fission of heavy nuclei and cause chain reaction. If all the neutrons produced during fission of uranium-235 produce further fission, then so much energy will be produced that it will not be controlled and leads to explosion called atom bomb. However, nuclear fission reaction can be controlled by using boron rods as boron can absorb neutrons. Nuclear fission reactions are done to generate electricity at nuclear power plants.

Nuclear Power Plant

Nuclear power plants use nuclear fission reaction to generate electricity and the fuel used for this purpose is uranium-235.

In a nuclear power plant fission reaction is carried out in a steel pressure vessel and inside is a nuclear reactor. In a nuclear reactor uranium-235 rods are inserted in graphite core. Graphite is called the moderator as it helps in slowing down the speed of neutrons so that a proper fission reaction takes place. In between the uranium-235 rods are placed boron rods as they help in absorbing excess neutrons and prevent nuclear fission reaction to of out of control. Boron rods are called control rods. The nuclear rods can be raised inside or pulled outside the reactor as the demand is. The nuclear reactor is enclosed in a concrete chamber which has thick wall so that it can absorb the nuclear radiations.

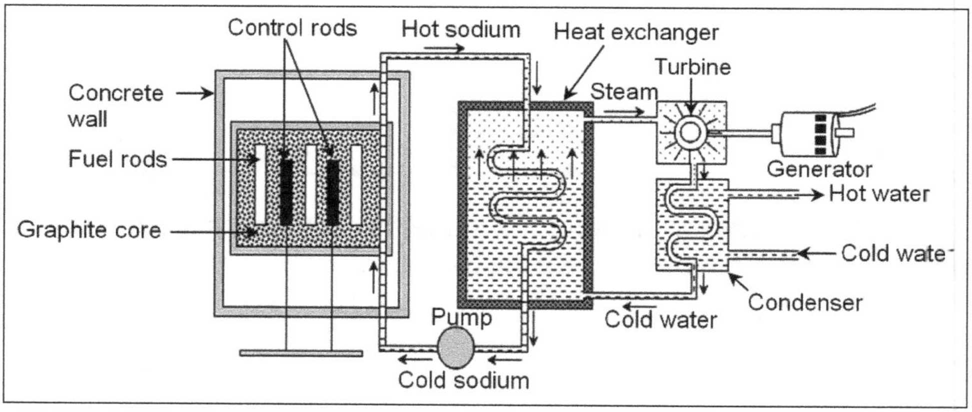

A Nuclear power plant or Atomic power plant.

Now the heat produced due to fission reaction in the reactor is cooled by using liquid sodium or carbon dioxide gas which also helps it transfer to heat exchanger. Here with the help of a coolant water is converted into steam. The steam produced is used to turn turbines and run generator.

Tremendous amount of heat energy is produced when controlled fission reaction takes place in nuclear reactor. That is why liquid sodium is pumped continuously through the pipes attached to the reactor. Sodium helps in absorbing the heat produced in the reactor. Then through pipes extremely hot sodium is passed through water in the heat exchanger. Water absorbs heat from hot sodium and boils to form steam. This steam is then passed at high pressure into turbine chamber having turbine. This steam then rotates the turbine which is further attached to its shaft and the generator. So, when turbine rotates, its shaft also rotates and drives generator. This generator helps in generating electricity.

The spent steam coming out of turbine chamber is passed through condenser which contains water and this water helps in cooling the steam. This steam then converts into water and through pipes is again sent to heat exchanger. The waste material produced in the nuclear fission reaction of uranium-235 is radioactive and extremely harmful for the environment.

Nuclear Bomb

Nuclear bomb is based on the nuclear fission reaction of uranium-235 and plutonium-239. The fission reaction is deliberately allowed to go out of control so as to produce large amount of energy in a very short while.

The atom bombs based on nuclear fission of uranium-235 ad plutonium-239 were dropped on Japanese cities of Hiroshima and Nagasaki in 1945 during the Second World War. This caused tremendous loss of human life.

Einstein's Mass-energy Relation

According to Einstein mass is equal to energy.

$$E = mc^2$$

Where,

- E is the amount of energy produced.
- M is mass destroyed.
- C is the speed of light in vacuum.

Since the speed of light is large so an extremely large amount of energy is produced even if small amount of mass is destroyed. Also, if mass is taken in kilograms (kg) and speed of light in meters per second (m/s) then energy will come in joules (J).

Therefore, if one kg of mass of any matter is destroyed in nuclear reaction than the amount of energy produced is:

- $E = mc^2$
- $E = 1 * (3 * 10^8)^2$
- $E = 9 * 10^{16}$ J

Energy Units for Expressing Nuclear Energy

The SI unit of energy released in nuclear reactions is electron volt (eV) or million electron volt (MeV). And,

1 electron volt = $1.602 * 10^{-19}$ joules

And,

1 million electron volt = $1.602 * 10^{-19} * 10^{6}$ joules

1 MeV = $1.602 * 10^{-13}$ J

Value of Atomic Mass Unit in Terms of Energy

Since absolute mass of atomic mass unit is $1.66 * 10^{-27}$ kg and the exact value of speed of light is $2.998 * 10^{8}$m/s. When we put these values in Einstein's equation we get,

1 atomic mass unit (u) = $1.492 * 10^{-10}$ J

also,

1 atomic mass unit (u) = 931 MeV

Nuclear Fusion

The meaning of fusion is to join or to combine. Therefore, the process in which two nuclei of light electrons are combined to form a heavy nucleus is nuclear fusion. In the process of nuclear fusion also tremendous amount of energy is released.

$$\underset{\text{Two deuterium atoms (Light atoms)}}{{}^{2}_{1}H} \xrightarrow{\text{Fusion}} \underset{\text{One helium atom (Heavy atom)}}{{}^{3}_{2}He} + \underset{\text{Neutron}}{{}^{1}_{0}n} + \text{Tremendous amount of energy}$$

The nuclei of atoms are positively charged and thus they repel each other. So in order to combine or fuse these two nuclei to form one heavy nucleus a lot of heat energy and high pressure is required. This shows that nuclear fusion is carried out by heating lighter atoms to an extremely high temperature at high pressure. Some mass is also lost in this process which gives tremendous amount of energy.

For example, when deuterium atoms are heated to an extremely high temperature under high pressure than two deuterium nuclei combine to form helium which has a heavy nucleus, a neutron is emitted and a lot of energy is liberated.

Nuclear fusion reaction is opposite of nuclear fission reaction. The energy produced in nuclear fusion reaction has not been controlled yet and is much more than nuclear fission reaction.

Hydrogen Bomb

Nuclear reactions which occur at extremely high temperature are called thermonuclear reactions. This reaction is used in producing hydrogen bomb which causes mass destruction. Isotopes of hydrogen, deuterium (^2H) and tritium (^3H), along with an element lithium-6 is used in making a hydrogen bomb. The explosion of hydrogen bomb is done by using an atom bomb. This is because when atom bomb is exploded then its fission reaction produced a lot of heat which raises the temperature of deuterium and tritium in a few microseconds. Thus the fusion reaction takes place and hydrogen bomb is exploded producing enormous energy. Hydrogen bomb causes destruction of life.

Advantages of Nuclear Energy

- It produced tremendous energy from a small amount of fuel (Uranium-235).
- There is no need to put the fuel again and again in nuclear reactor. One the fuel (Uranium-235) is put in the reactor it can function for two to three years at a stretch.
- It does not produce gases like carbon dioxide or sulphur dioxide.

Disadvantages of Nuclear Energy

- The waste products of nuclear reactors are radioactive and keep on emitting harmful radiations.
- Risk of an accident in nuclear reactors which may cause leakage of radioactive material.
- The availability of fuel uranium is limited.
- High installation cost of nuclear power plant.

Compound Nucleus Reactions

A projectile and a target nucleus are within the range of nuclear forces for the time allowing for a large number of interactions between nucleons. These type of reactions are called the compound nucleus reactions.

In fact, there is always some non-direct (multiple internuclear interaction) component in all reactions, but the direct reactions have this component limited.

What are the Compound Nucleus and the Nuclear Resonance?

There is no difference between the compound nucleus and the nuclear resonance.

The compound nucleus is the intermediate state formed in a compound nucleus reaction. It is normally one of the excited states of the nucleus formed by the combination of the incident particle and target nucleus. If a target nucleus X is bombarded with particles a, it is sometimes observed that the ensuing nuclear reaction takes place with appreciable probability only if the energy of the particle a is in the neighborhood of certain definite energy values. These energy values are referred to as resonance energies. The compound nuclei of these certain energies are reffered to as nuclear resonances. Resonances are usually found only at relatively low energies of the projectile. The widths of the resonances increase in general with increasing energies. At higher energies the widths may reach the order of the distances between resonances and then no resonances can be observed. The narrowest resonances are usually the compound states of heavy nuclei (such as fissionable nuclei) and thermal neutrons (usually in (n,γ) capture reactions). The observation of resonances is by no means restricted to neutron nuclear reactions.

The compound nucleus model (idea of compound nucleus formation) was introduced by Danish physicist Niels Bohr in 1936. This model assumes that incident particle and the target nucleus become indistinguishable after the collision and together constitute the particular excited state of nucleus – the compound nucleus. To become indistinguishable the projectile has to suffer collisions with constituent nucleons of the target nucleus until it has lost its incident energy. In fact many so these collisions lead to a complete thermal equilibrium inside the compound nucleus. The compound nucleus is excited by both the kinetic energy of the projectile and by the binding nuclear energy.

This compound system is a relatively long-lived intermediate state of particle-target composite system and from the definition, the compound nucleus must live for at least several times longer than is the time of transit of an incident particle across the nucleus (~10^{-22} s). The time scale of compound nucleus reactions is of the order of 10^{-18} s – 10^{-16} s, but lifetimes as long as 10^{-14} s have been also observed.

Very important feature and a direct consequence of the thermal equilibrium inside a compound nucleus is the fact the mode of decay of compound nucleus does not depend on the way the compound nucleus is formed. The large number of collisions between nucleons leads to the loss of the information on the entrance channel from the system. The decay mechanism (exit channel) that dominates the decay of C* is determined by the excitation energy in C* and by the law of probability.

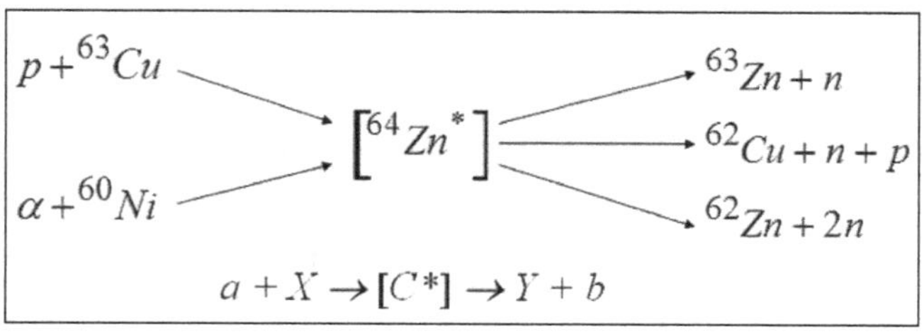

These reactions can be considered as a two-stage processes:

- The first stage is the formation of a compound nucleus expressed by $\sigma_{a+X \to C^*}$
- The second stage is the decay of a compound nucleus expressed by $P_{C^* \to b+Y}$
- The result cross-section of certain reaction a+X→[C*]→b+Y is given by $\sigma_{(a,b)} = \sigma_{a+X \to C^*} \cdot P_{C^* \to b+Y}$

$$ {}_0^1n + {}_{92}^{235}U \to [{}_{92}^{236}U] \xrightarrow{85\%} fission(e.g. {}_{56}^{139}Ba + {}_{36}^{94}Kr + 3{}_0^1n) $$

$$ {}_0^1n + {}_{92}^{235}U \to [{}_{92}^{236}U] \xrightarrow{15\%} {}_{92}^{236}U + \gamma $$

Absorption reaction of fissile 235U. The uncertainty of the exit channel is caused by “loss of memory” of resonance [236U].

Resonances

For the compound nucleus peaks in the cross-section are typical. Each peak is manifesting a particular compound state of nucleus. These peaks and the associated compound nuclei are usually called resonances. The behaviour of the cross-section between two resonances is usually strongly affected by the effect of nearby resonances.

Resonances (particular compound states) are mostly created in neutron nuclear reactions, but it is by no means restricted to neutron nuclear reactions. The formation of resonances is caused by the quantum nature of nuclear forces. Each nuclear reaction is a transition between different quantum discrete states or energy levels. The discrete nature of energy transitions plays a key role. If the energy of the projectile (the sum of the Q value and the kinetic energy of the projectile) and the energy of target nucleus is equal to a compound nucleus at one of the excitation states, a resonance can be created and peak occurs in the cross section. For light nucleus, the allowable state density in this energy region is much lower and the “distance” between states is higher. For heavy nuclei, such as ^{238}U, we can observe large resonance region in the neutron absorption cross-section.

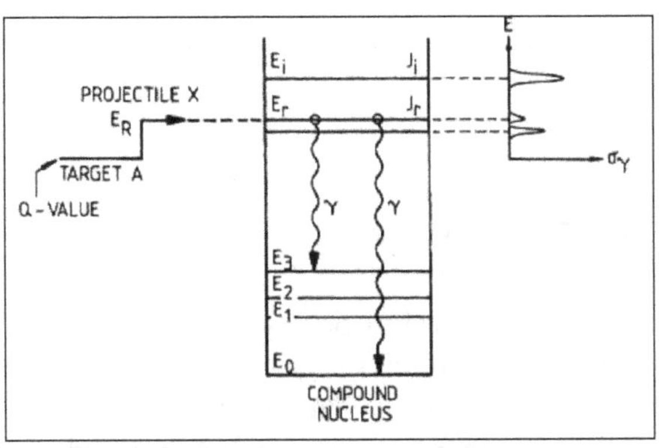

Energy levels of compound state. For neutron absorption reaction on 238U the first resonance E1 corresponds to the excitation energy of 6.67eV. E0 is a base state of 239U.

It is obvious the compound states (resonances) are observed at low excitation energies. This is due to the fact, the energy gap between the states is large. At high excitation energy, the gap between two compound states is very small and the widths of resonances may reach the order of the distances between resonances. Therefore at high energies no resonances can be observed and the cross section in this energy region is continuous and smooth.

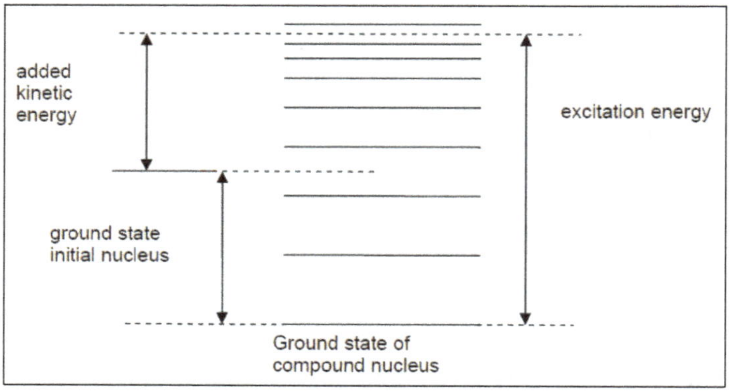

The position of the energy levels during the formation of a compound nucleus. Ground state and energy states.

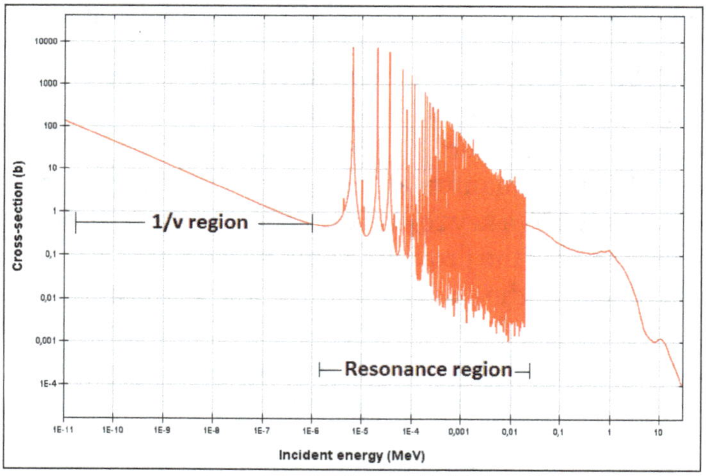

Region of resonances of ^{238}U nuclei.

Nuclear Transmutation

Nuclear transmutation is the conversion of one chemical element or isotope into another. In other words, atoms of one element can be changed into atoms of another element by transmutation. This occurs either through nuclear reactions in which an outside particle reacts with a nucleus, which can be supplied by a particle accelerator, or through radioactive decay, where no outside particle is needed.

The term transmutation dates back to alchemy. Alchemists pursued the philosopher's stone, capable of the transformation of base metals into gold. The impossibility of the metallic transmutation had been debated amongst alchemists, philosophers, and scientists since the Middle Ages. In the 18th century, Antoine Lavoisier replaced the alchemical theory of elements with the modern theory of chemical elements, and later John Dalton further developed the notion of atoms to explain various chemical processes. The disintegration of atoms is a distinct process involving much greater energies than could be achieved by alchemists.

Nuclear transmutation was first consciously applied to modern physics by Frederick Soddy when he, along with Ernest Rutherford, discovered that radioactive thorium was converting itself into radium in 1901.

Particle Accelerators

There are two basic classes of accelerators: electrostatic and oscillating field accelerators. Electrostatic accelerators use static electric fields to accelerate particles. A small-scale example of this class is the cathode ray tube in an ordinary, old television set. Other examples are the Cockcroft–Walton generator and the Van de Graaf generator. The achievable kinetic energy for particles in these devices is limited by electrical breakdown. Oscillating field accelerators, on the other hand, use radio frequency electromagnetic fields to circumvent the breakdown problem. This class, development for which started in the 1920's, is the basis for all modern accelerator concepts and large-scale facilities. Rolf Widerøe, Gustav Ising, Leó Szilárd, Donald Kerst and Ernest Lawrence are considered pioneers of the field, conceiving and building the first operational linear particle accelerator, the betatron, and the cyclotron.

Since colliders can give evidence on the structure of the subatomic world, accelerators were commonly referred to as atom smashers in the 20th century. Despite the fact that most accelerators (with the exception of ion facilities) actually propel subatomic particles, the term persists in popular usage when referring to particle accelerators in general.

Fermi National Accelerator Laboratory: Aerial photo of the Tevatron at Fermilab, which resembles a figure eight. The main accelerator is the ring above; the one below (about half the diameter, despite appearances) is for preliminary acceleration, beam cooling and storage, etc.

Predicting the products of transmutation is like predicting the products of radioactive decay. It is important to make sure that the total atomic mass and atomic numbers of both sides of the equation remain equal.

Transuranium Elements

Transuranium elements are those beyond uranium, none of which is stable because of radioactive decomposition.

In chemistry, transuranium elements, also known as transuranic elements, are the chemical elements with atomic numbers greater than 92, which is the atomic number of uranium. None of these elements is stable and each of them decays radioactively into other elements.

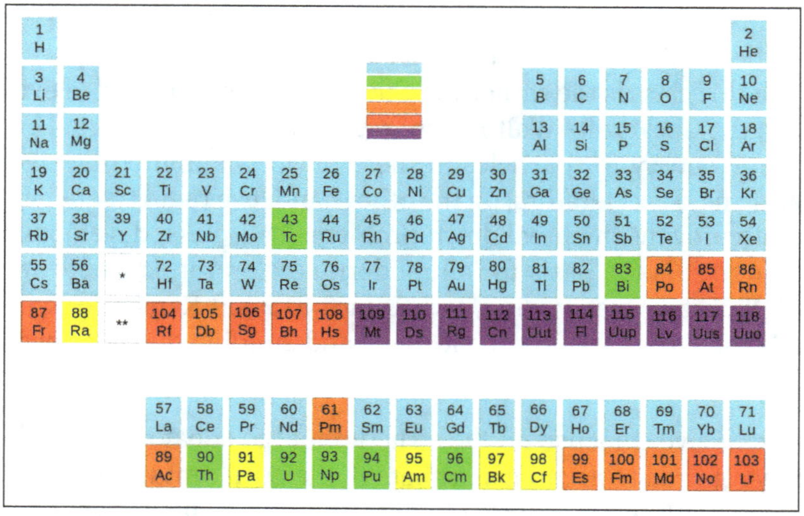

Periodic table radioactivity: Blue – Elements that contain at least one stable isotope. Green – Radioactive elements: the most stable isotope is very long-lived, with s half-life of over four million years. Yellow – Radioactive elements: the most stable isotope has a half-life between 800 and 34.000 years. Orange – Radioactive elements: the most stable isotope has a half-life between one day and 103 years. Red – Highly radioactive elements: the most stable isotope has a half-life between several minutes and one day. Purple – Extremely radioactive elements: the most stable isotope has a half-life less than several minutes. Very little is known about these elements due to their extreme instability and radioactivity.

Transuranium Elements in Nature

All of the elements with atomic numbers 1 to 92 can be found in nature and have stable or very long half-life isotopes. They can also be created as common products of the decay of uranium and thorium.

However, all of the elements with higher atomic numbers have been first discovered in the laboratory. Each of these elements is radioactive, with a half-life much shorter than the age of the Earth. So, if any atoms of these elements were ever present at the Earth's formation, they have long since decayed.

Producing Transuranium Elements

Transuranium elements that can be found on Earth now are artificially-generated, synthetic elements made via nuclear reactors or particle accelerators. The half-lives of these elements show a general trend of decreasing as atomic numbers increase. However, there are exceptions, including dubnium and several isotopes of curium. Further anomalous elements in this series have been predicted by Glenn T. Seaborg. They are categorized as the "island of stability."

Heavy transuranic elements are difficult and expensive to produce. Their prices go up rapidly with atomic number. As of 2008, weapons-grade plutonium cost around $4,000 per gram and californium cost $60,000,000 per gram. Due to production difficulties, none of the elements beyond californium has industrial applications, and of them, only einsteinium has ever been produced in macroscopic quantities.

Neutron Elastic Scattering

A neutron scattering reaction occurs when a target nucleus emits a single neutron after a neutron-nucleus interaction. In an elastic scattering reaction between a neutron and a target nucleus, there is no energy transferred into nuclear excitation. The elastic scattering conserves both momentum and kinetic energy of the "system". It may be modeled as a billiard ball collision between a neutron and a nucleus.

There is usually some transfer of kinetic energy from the incident neutron to the target nucleus. The target nucleus gains the exact amount of kinetic energy that the neutron loses. This interaction can take place via compound nucleus formation, but, in case of elastic scattering, a neutron emission returns the compound nucleus to the ground state of the original nucleus. Therefore the initial and final neutrons do not need to be necessarily the same. The elastic scattering can occur by way of two interaction mechanisms:

- Potential scattering: In potential scattering, the neutron and the nucleus interact without neutron absorption and the formation of a compound nucleus. In fact, the incident neutron does not necessarily have to "touch" the nucleus and the neutron is scattered by the short range nuclear forces when it approaches close enough to the nucleus. Potential scattering occurs with incident neutrons that have an energy of up to about 1 MeV. It may be modeled as a billiard ball collision between a neutron and a nucleus.

- Compound-elastic scattering: In some cases, if the kinetic energy of an incident neutron just right to form a resonance, the neutron may be absorbed and a compound nucleus may be formed. This interaction is more unusual and is also known as resonance elastic scattering. Due to formation of the compound nucleus, initial and final neutron are not the same.

Conservation of Momentum and Kinetic Energy

$$(m_n.v_{n,i})+(m_T.v_{T,i})=(m_n.v_{n,f})+(m_T.v_{T,f})$$
$$(1/2m_n.v_{n,i}^2)+(1/2m_T.v_{T,i}^2)=(1/2m_n.v_{n,f}^2)+(1/2m_T.v_{T,f}^2)$$

where:

m_n = mass of the neutron

m_T = mass of the target nucleus T

$v_{n,i}$ = initial neutron speed

$v_{n,f}$ = final neutron speed

$v_{T,i}$ = initial target speed

$v_{T,f}$ = final target speed

Key Characteristics of Elastic Scattering

- Elastic scattering is the most important process for slowing down neutrons.
- Total kinetic energy of the system is conserved in elastic scattering.
- In this process, energy lost by the neutron is transferred to the recoiling nucleus.
- Maximum energy transfer is occurred with an head-on collision.
- Kinetic energy of the recoiled nucleus depends on the recoiled angle φ of the nucleus.
- Elastic scattering cross-sections for light elements are more or less independent of neutron energy up to 1 MeV.
- For intermediate and heavy elements, the elastic cross-section is constant at low energy with some specifics at higher energy.
- A good approximation is, σ_s = const, for all elements, that are of importance.
- At low energy, σ_s can be described by the one-level Breit-Wigner formula.

- Nearly all elements have scattering cross-sections in the range of 2 to 20 barns.
- The important exception is for water and heavy water.
- If the kinetic energy of an incident neutron is large compared with the chemical binding energy of the atoms in a molecule, the chemical bound can be ignored.
- If the kinetic energy of an incident neutron is of the order or less than the chemical binding energy, the cross-section of the molecule is not equal to the sum of cross-sections of its individual nuclei.
- Scattering of slow neutrons by molecules is greater than by free nuclei.
- Therefore one nucleus microscopic cross-sections do not describe the process correctly, while the macroscopic cross-section (Σs) has a precise meaning.

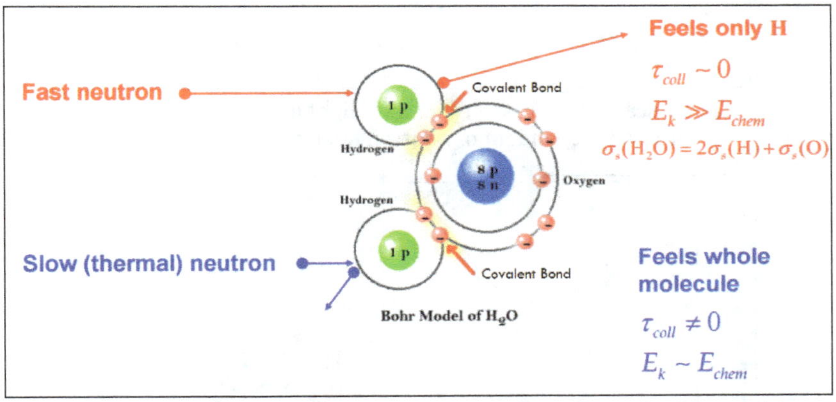

Scattering of slow neutrons by molecules is greater than by free nuclei.

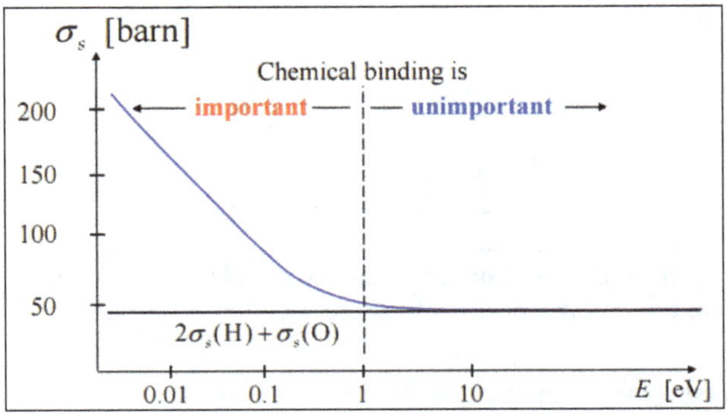

If the kinetic energy of an incident neutron is large compared with the chemical binding energy of the atoms in a molecule, the chemical bound can be ignored. If the kinetic energy of an incident neutron is of the order or less than the chemical binding energy, the cross-section of the molecule is not equal to the sum of cross-sections of its individual nuclei.

Elastic Scattering Cross-section

To be an effective moderator, the probability of elastic reaction between neutron and the nucleus must be high. In terms of cross-sections, the elastic scattering cross section of a moderator's nucleus must be high.

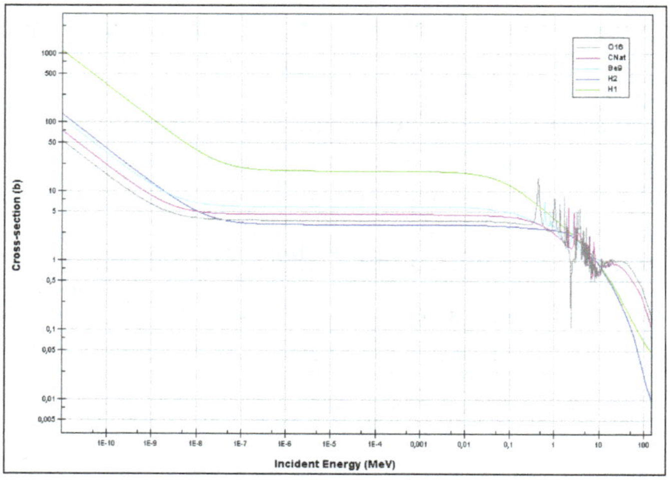

Elastic scattering cross-sections for light elements are more or less independent of neutron energy up to 1 MeV.

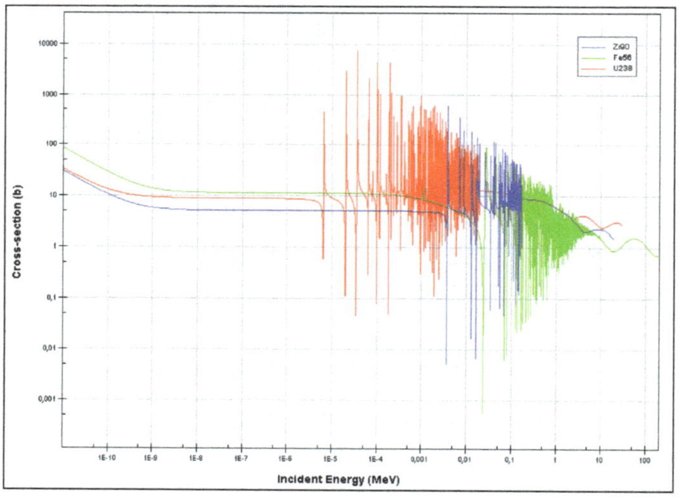

For intermediate and heavy elements, the elastic cross-section is constant at low energy with some specifics at higher energy.

Elastic Scattering and Neutron Moderators

As can be seen, a high elastic scattering cross-section is important, but does not describe comprehensively capabilities of moderators. In order to describe capabilities of a material to slow down neutrons, three new material variables must be defined:

- The Average Logarithmic Energy Decrement (ξ)

- The Macroscopic Slowing Down Power (MSDP)
- The Moderating Ratio (MR)

Key properties of neutron moderators:

- High cross-section for neutron scattering
- High energy loss per collision
- Low cross-section for absorption
- High melting and boiling point
- High thermal conductivity
- High specific heat capacity
- Low viscosity
- Low activity
- Low corrosive
- Cheap

Average Logarithmic Energy Decrement

During the scattering reaction, a fraction of the neutron's kinetic energy is transferred to the nucleus. Using the laws of conservation of momentum and energy and the analogy of collisions of billiard balls for elastic scattering, it is possible to derive the following equation for the mass of target or moderator nucleus (M), energy of incident neutron (E_i) and the energy of scattered neutron (E_s).

$$E_s = \left(\frac{M - m_n}{M + m_n}\right)^2 E_i, \quad E_s = \left(\frac{A-1}{A+1}\right)^2 E_i$$

Where A is the atomic mass number.

In case of the hydrogen (A = 1) as the target nucleus, the incident neutron can be completely stopped. But this works when the direction of the neutron is completely reversed (i.e. scattered at 180°). In reality, the direction of scattering ranges from 0 to 180 ° and the energy transferred also ranges from 0% to maximum. Therefore, the average energy of scattered neutron is taken as the average of energies with scattering angle 0 and 180°.

Moreover, it is useful to work with logarithmic quantities and therefore one defines the logarithmic energy decrement per collision (ξ) as a key material constant describing energy transfers during a neutron slowing down. ξ is not dependent on energy, only on A and is defined as follows:

$$\xi = \overline{\ln E' - \ln E} = \int_{aE'}^{E'} \ln \frac{E'}{E} p(E' \to E) dE = 1 + \frac{\alpha}{1-\alpha} \ln \alpha$$

Where,

$$\left(\frac{A-1}{A+1}\right)^2 = \alpha$$

For heavy target nuclei, ξ may be approximated by following formula:

$$\xi \simeq \frac{2}{A + 2/3} \quad \text{for } A > 10$$

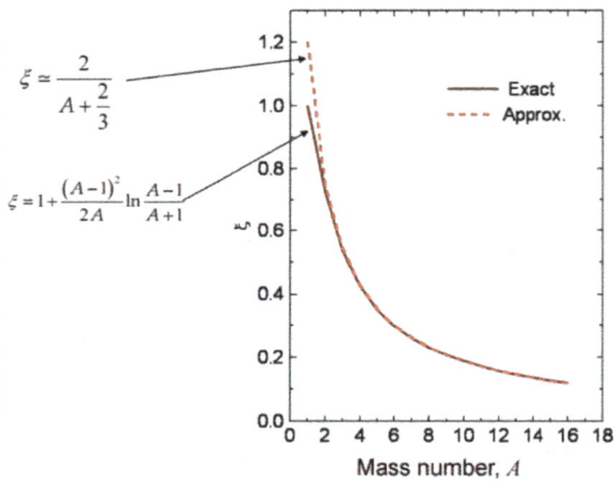

From these equations it is easy to determine the number of collisions required to slow down a neutron from, for example from 2 MeV to 1 eV.

Example: Determine the number of collisions required for thermalization for the 2 MeV neutron in the carbon.

$\xi_{CARBON} = 0.158$

$N(2MeV \to 1eV) = \ln 2 \cdot 10^6/\xi = 14.5/0.158 = 92$

Table: Average logarithmic energy decrement for some elements.

Element	A	î	Collisions 2 MeV → 1eV
H	1	1	15
D	2	0.750	20
H_2O	-	0.920	16

D_2O	-	0.509	29
He	4	0.425	34
Be	9	0.207	70
C	12	0.158	92
O	16	0.120	121
Na	23	0.084	172
Fe	56	0.035	414
^{238}U	238	0.008	1812

For a mixture of isotopes:

$$\bar{\xi} = \frac{Sum(\xi_i \cdot \sum_{si})}{Sum(\sum_{si})}$$

Macroscopic Slowing Down Power

We have defined the probability of elastic scattering reaction, we have defined the average energy loss during the reaction. The product of these variables (the logarithmic energy decrement and the macroscopic cross section for scattering in the material) is the macroscopic slowing down power (MSDP).

$$MSDP = \xi \,.\, \Sigma_s$$

The MSDP describes the ability of a given material to slow down neutrons and indicates how rapidly a neutron will slow down in the material, but it does not fully reflect the effectiveness of the material as a moderator. In fact, the material with high MSDP can slow down neutrons with high efficiency, but it can be a poor moderator because of its high probability of absorbing neutrons. It is typical, for example, for boron, which has a high slowing down power but is absolutely inappropriate as a moderator.

The most complete measure of the effectiveness of a moderator is the Moderating Ratio (MR), where:

$$MR = \xi \,.\, \Sigma_s/\Sigma_a$$

Table: Macroscopic slowing down power MSDP for some materials.

Element	A	ξ	Collisions 2MeV)→1eV	MSDP	MR
H_2O	-	0.920	16	1.35	71
D_2O	-	0.509	29	0.176	5670
Be	9	0.207	70	0.158	143
C	12	0.158	92	0.060	192

Moderating Ratio

The moderating ratio or moderator quality is the most complete measure of the effectiveness of a moderator because it takes into account also the absorption effects. When absorption effects are high, most of the neutrons will be absorbed by moderator, leading to lower moderation or lower availability of thermal neutrons.

Therefore a higher ratio of MSDP to absorbtion cross sections $\xi . \Sigma_s/\Sigma_a$ is desirable for effective moderation. This ratio is called the moderating ratio – MR and can be used as a criterion for comparison of different moderators.

Examples:

- Light water has the highest ξ and σ_s among the moderators (resulting in the highest MSDP) shown in the table, but its moderating ratio is low due to its relatively higher absorption cross section.
- On the other hand, heavy water has lower ξ and σs, but it has the highest moderating ratio owing to its lowest neutron absorption cross-section.
- Graphite has much heavier nuclei than hydrogen in water, despite the fact graphite has much lower ξ and σ_s, it is better moderator than light water due to its lower absorption cross-section compared to that of light water.

Table: Moderating ratios for some materials.

Element	A	ξ	Collisions 2MeV⟩→1eV	MR
H_2O	-	0.920	16	71
D_2O	-	0.509	29	5670
Be	9	0.207	70	143
C	12	0.158	92	192
238U	238	0.008	1812	0.0092

Neutron Inelastic Scattering

A neutron scattering reaction occurs when a target nucleus emits a single neutron after a neutron-nucleus interaction. In an elastic scattering reaction between a neutron and a target nucleus, there is no energy transferred into nuclear excitation.

Besides, in an inelastic scattering reaction between a neutron and a target nucleus some energy of the incident neutron is absorbed to the recoiling nucleus and the nucleus remains in the excited state. Thus while momentum is conserved in an inelastic collision, kinetic energy of the “system” is not conserved.

In inelastic neutron scattering, the incident neutron is absorbed by the target nucleus and the interaction take place via compound nucleus formation. The compound nucleus emits then a neutron of lower kinetic energy which leaves the original nucleus in an excited state. The nucleus gives up excitation energy by emitting one or more gamma rays to reach its ground state. In comparison to elastic scattering reactions, it is not easy to write an expression for the average energy loss of inelastic scattering reactions because it depends on the energy levels within the target nucleus.

Inelastic scattering occurs above a threshold energy. This threshold energy is higher than the energy the first excited state of target nucleus (due to the laws of conservation) and it is given by following formula:

$$E_t = ((A+1)/A)^* \varepsilon_1$$

where E_t is known as the inelastic threshold energy and ε_1 is the energy of the first excited state.

In general, the energy of the first excited state of nuclei decreases with increasing mass number. Therefore an inelastic scattering plays an important role in slowing down neutrons especially at high energies and by heavy nuclei. In calculations of a reactor inelastic scattering by light nuclei (such as oxygen with E_t = 6.42 MeV) can ordinarily be ignored. Moreover, the hydrogen nucleus does not have excited states, so only an elastic scattering reaction can occur in that case.

On the other hand an inelastic scattering by uranium nuclei (especially by ^{238}U) must be included. ^{238}U is a major component of nuclear fuel in commercial power reactors, therefore its scattering data are one of the most important data in the neutron transport calculations in the reactor core.

Key Characteristics of Inelastic Scattering

- During an inelastic scattering the neutron is absorbed and then re-emitted.
- The reaction occurs via compound nucleus.
- While momentum is conserved in an inelastic collision, kinetic energy of the "system" is not conserved.
- Some energy of the incident neutron is absorbed to the recoiling nucleus and the nucleus remains in the excited state.
- The nucleus gives up excitation energy by emitting one or more gamma rays.
- General notation: A(n, n')A* or A(n, 2n')B; Example: 14O(n, n')14O*.
- Inelastic scattering is a threshold reaction and occurs above a threshold energy.

- Inelastic scattering cross section is relatively small for light nuclei.
- For hydrogen nucleus, inelastic scattering does not occur, because it does not have excited states.
- Inelastic scattering plays an important role in slowing down neutrons especially at high energies and by heavy nuclei (e.g. ^{238}U).
- Inelastic scattering may be significant for heterogeneous reactors with highly enriched fuel (e.g. in fast neutron reactors).

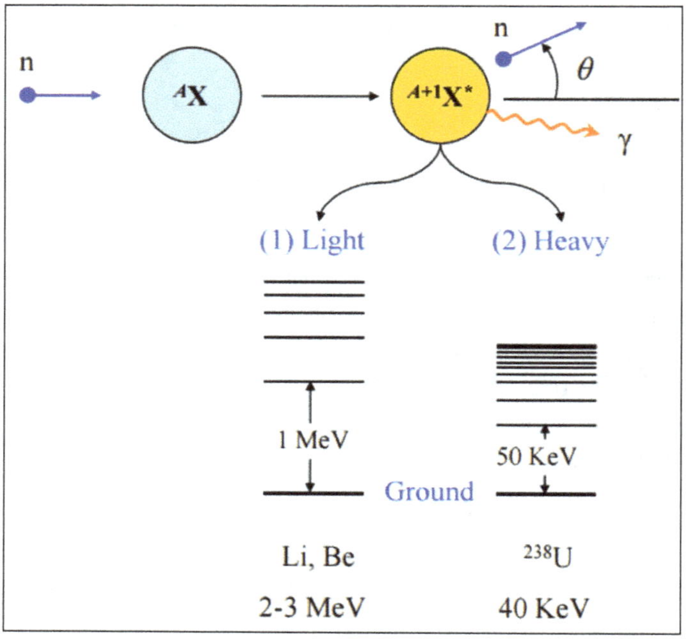

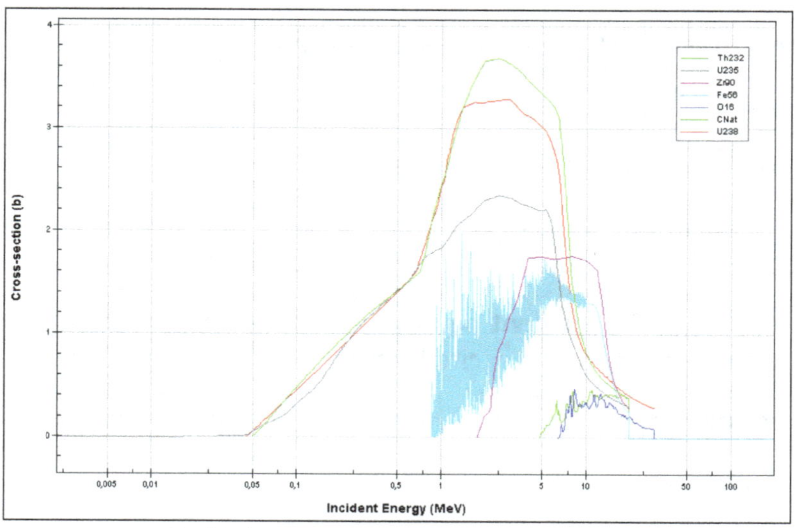

Inelastic neutron scattering cross sections.

Neutron Capture

Neutron capture is a type of nuclear reaction in which a target nucleus absorbs a neutron (uncharged particle), then emits a discrete quantity of electromagnetic energy (gamma-ray photon). The target nucleus and the product nucleus are isotopes, or forms of the same element. Thus phosphorus-31, on undergoing neutron capture, becomes phosphorus-32. The heavier isotope that results may be radioactive, so that neutron capture, which occurs with almost any nucleus, is a common way of producing radioactive isotopes.

Neutron capture is also named neutron-gamma, or (η,γ), reaction from the bombarding particle (η for neutron) and the emitted particle (γ for gamma-ray photon) and sometimes called neutron radiative capture because of the prompt emission of only electromagnetic radiation. Among the natural elements, boron, cadmium, and gadolinium are the best absorbers of slow neutrons by the capture process.

References

- Nuclear-reactions, nuclear-power: nuclear-power.net, Retrieved 2 January, 2019
- Nuclear-equations: opentextbc.ca, Retrieved 22 May, 2019
- Energetics-of-Nuclear-Reactions, Nuclear-Chemistry: libretexts.org, Retrieved 13 February, 2019
- Photonuclear-Reaction: thefreedictionary.com, Retrieved 28 May, 2019
- Nuclear-fission-and-nuclear-fusion: jagranjosh.com, Retrieved 17 July, 2019
- Compound-nucleus-reactions, neutron-nuclear-reactions, reactor-physics: nuclear-power.net, Retrieved 23 April, 2019
- Nuclear-transmutation: lumenlearning.com, Retrieved 26 August, 2019
- Neutron-elastic-scattering, neutron-nuclear-reactions, nuclear-engineering, reactor-physics: nuclear-power.net, Retrieved 15 March, 2019
- Neutron-capture, science: britannica.com, Retrieved 14 June, 2019

5

Chemistry of Nuclear Power

Nuclear chemistry plays an integral role in the synthesis of thorium and uranium for use in nuclear reactors. It is also involved in the fabrication of nuclear fuel as well as in the treatment of radioactive waste. The chapter closely examines these key applications of nuclear chemistry in the field of nuclear power to provide an extensive understanding of the subject.

Nuclear Reactors

The energy released from nuclear fission can be harnessed to make electricity with a nuclear reactor. A nuclear reactor is a piece of equipment where nuclear chain reactions can be controlled and sustained. The reactors use nuclear fuel, most commonly uranium-235 and plutonium-239. The amount of free energy in nuclear fuels is far greater than the energy in a similar amount of other fuels such as gasoline. In many countries, nuclear power is seen as an environmentally friendly alternative to fossil fuels, which are non-renewable and release large amounts of greenhouse gases. However, nuclear reactors produce nuclear waste containing radioactive elements.

The Chain Reaction

When a large, fissile atomic nucleus such as uranium-235 or plutonium-239 absorbs a neutron, it may undergo nuclear fission. The nucleus splits into two or more lighter nuclei, releasing kinetic energy, gamma radiation, and free neutrons. A portion of these neutrons may later be absorbed by other fissile atoms and trigger further fission events, which release more neutrons, and so on. This is known as a nuclear chain reaction.

A possible nuclear fission chain reaction: In the first step, a uranium-235 atom absorbs a neutron, and splits into two new atoms (fission fragments), releasing three new neutrons and a large amount of binding energy. In the second step, one of those neutrons is absorbed by an atom of uranium-238, and does not continue the reaction. Another neutron leaves the system without being absorbed. However, one neutron does collide with an atom of uranium-235, which then splits and releases two neutrons and more binding energy. In the third step, both of those neutrons collide with uranium-235 atoms, each of which splits and releases a few neutrons, which can then continue the reaction.

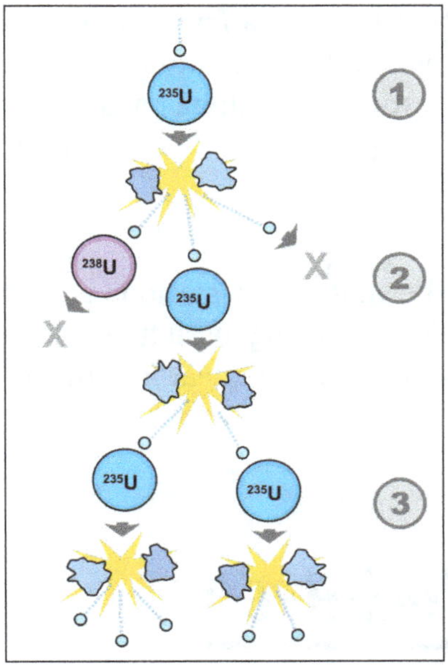

Nuclear chain reaction.

This chain reaction can be controlled using neutron poisons and neutron moderators to change the portion of neutrons that can cause more fissions. A neutron moderator works to reduce a newly produced neutron's kinetic energy from several MeV to thermal energies of less than one eV, making them more likely to induce further fission.

Nuclear reactors generally have automatic and manual systems to shut the fission reaction down if unsafe conditions are detected. The amount and nature of neutron moderation affects reactor controllability and safety. Since moderators both slow and absorb neutrons, there is an optimum amount of moderator to include in a given geometry of reactor core.

Criticality

In a nuclear reactor, the neutron population at any instant is a function of the rate of neutron production and the rate of neutron loss. When a reactor's neutron population remains steady from one generation to the next by creating as many new neutrons as are lost, the fission chain reaction is self-sustaining and the reactor's condition is referred to as "critical." When the reactor's neutron production exceeds losses, characterized by increasing power level, it is considered "supercritical." When losses dominate, it is considered "subcritical" and exhibits decreasing power.

The mere fact that an assembly is supercritical does not guarantee that it contains any free neutrons at all. At least one neutron is required to "strike" a chain reaction, and if the spontaneous fission rate is sufficiently low, it may take a long time before a chance neutron encounter starts a chain reaction—even if the reactor is supercritical. In ^{235}U reactors, this time might be a long as many minutes. Most nuclear reactors include a "starter" neutron source that ensures a few free neutrons in the reactor core, so a chain reaction will occur immediately when the core is made critical. A common type of startup neutron source is a mixture of an alpha particle emitter such as ^{241}Am (americium-241) with a lightweight isotope such as ^{9}Be (beryllium-9).

Generating Electricity

Just as many conventional thermal power stations generate electricity by harnessing the thermal energy released from burning fossil fuels, nuclear power plants convert the energy released from nuclear fission. The heat is removed from the reactor core by a cooling system that generates steam. The steam drives a turbine which runs a generator to produce electricity.

Nuclear Chain Reaction

Nuclear chain reactions are reactions where nuclear energy is obtained, generally through nuclear fission. These chain reactions are what provide nuclear power plants with the energy that is then turned into electricity for use by people. In these reactions, neutrons generated by the fission process continue on to initiate fission in other atoms. These reactions generally occur with heavier isotopes such as uranium-235 where there is a continuous release and absorption of neutrons. Figure below is a visual representation of what a nuclear fission chain reaction looks like. If at least one neutron from each fission strikes another U-235 nucleus and initiates fission, the chain reaction is sustained and it is said to be critical.

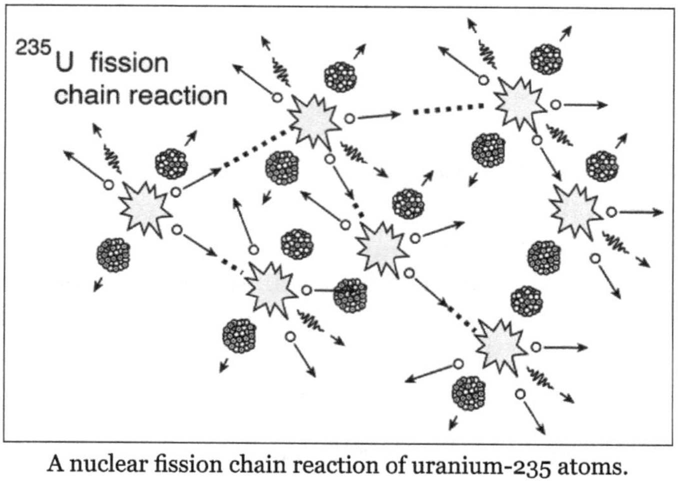

A nuclear fission chain reaction of uranium-235 atoms.

The mass of uranium-235 that is required to produce a reaction that is self-sustaining is said to be the "critical mass". This critical chain reaction can be accomplished at relatively low uranium-235 concentrations if the neutrons are moderated to lower their speed. A moderator is necessary because the probability for fission is greater with slow neutrons. The general equation for a nuclear chain reaction of uranium-235 is:

$$Uranium - 235 + {}_0^1 n \rightarrow\sim 3{}_0^1 n + \text{energy}$$

If for every one neutron inputted into the equation releases two or three more neutrons, then the number of fission events increases dramatically each generation. However, in reality not all of the released neutrons actually cause more fissions. Only 1.1 neutron per reaction actually goes on to cause more fissions and continue the chain, however the number of fission events still grows quickly. The process of a nuclear chain reaction releases large amounts of energy, but this energy can be utilized in different ways. On average, there is about 200 MeV of energy released during fission. To put this into context, burning coal provides only a couple eV, while 200 MeV is equal to 200 million electron volts. The difference in these energies is enormous. In nuclear reactors, the reaction is moderated and progresses at a slow pace to release its energy over a period of time so it can be harnessed and used for peaceful purposes. An atomic bomb utilizes this fission chain reaction as well, however it is designed to release its energy all at once—which is much more damaging. In either case, the release of the energy is controlled, but the time period taken to release the energy differs.

Nuclear Fuel Cycle

The nuclear fuel cycle is the series of industrial processes which involve the production of electricity from uranium in nuclear power reactors.

Uranium is a relatively common element that is found throughout the world. It is mined in a number of countries and must be processed before it can be used as fuel for a nuclear reactor.

Fuel removed from a reactor, after it has reached the end of its useful life, can be reprocessed so that most is recycled for new fuel.

The various activities associated with the production of electricity from nuclear reactions are referred to collectively as the nuclear fuel cycle. The nuclear fuel cycle starts with the mining of uranium and ends with the disposal of nuclear waste. With the reprocessing of used fuel as an option for nuclear energy, the stages form a true cycle.

To prepare uranium for use in a nuclear reactor, it undergoes the steps of mining and milling, conversion, enrichment and fuel fabrication. These steps make up the 'front end' of the nuclear fuel cycle.

After uranium has spent about three years in a reactor to produce electricity, the used fuel may undergo a further series of steps including temporary storage, reprocessing, and recycling before wastes are disposed. Collectively these steps are known as the 'back end' of the fuel cycle.

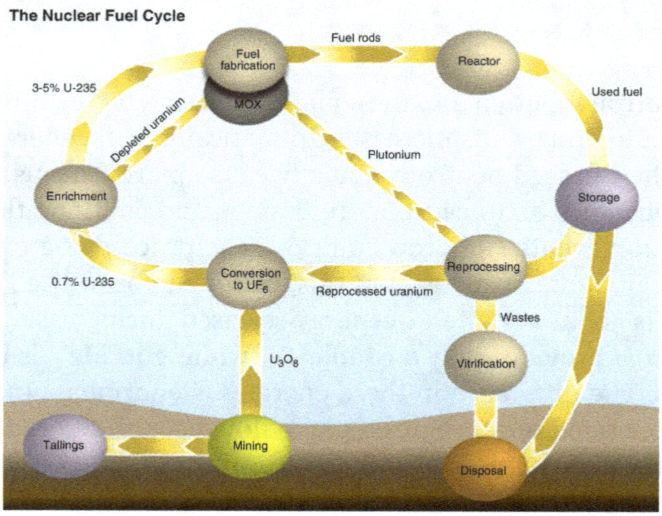

Uranium

Uranium is a slightly radioactive metal that occurs throughout the Earth's crust. It is about 500 times more abundant than gold and about as common as tin. It is present in most rocks and soils as well as in many rivers and in sea water. It is, for example, found in concentrations of about four parts per million (ppm) in granite, which makes up 60% of the Earth's crust. In fertilisers, uranium concentration can be as high as 400 ppm (0.04%), and some coal deposits contain uranium at concentrations greater than 100 ppm (0.01%). Most of the radioactivity associated with uranium in nature is in fact due to other minerals derived from it by radioactive decay processes, and which are left behind in mining and milling.

There are a number of areas around the world where the concentration of uranium in the ground is sufficiently high that extraction of it for use as nuclear fuel is economically feasible. Such concentrations are called ore.

Uranium Milling

Milling, which is generally carried out close to a uranium mine, extracts the uranium from the ore (or ISL leachate). Most mining facilities include a mill, although where mines are close together, one mill may process the ore from several mines. Milling produces a uranium oxide concentrate which is shipped from the mill. It is sometimes referred to as 'yellowcake' and generally contains more than 80% uranium. The original ore may contain as little as 0.1% uranium, or even less.

In a mill, the ore is crushed and ground to a fine slurry which is leached in sulfuric acid (or sometimes a strong alkaline solution) to allow the separation of uranium from the waste rock. It is then recovered from solution and precipitated as uranium oxide (U_3O_8) concentrate. After drying and usually heating it is packed in 200-litre drums as a concentrate, sometimes referred to as 'yellowcake' (though it is usually khaki).

U_3O_8 is the uranium product which is sold. About 200 tonnes is required to keep a large (1000 MWe) nuclear power reactor generating electricity for one year.

The remainder of the ore, containing most of the radioactivity and nearly all the rock material, becomes tailings, which are emplaced in engineered facilities near the mine (often in a mined out pit). Tailings need to be isolated from the environment because they contain long-lived radioactive materials in low concentrations and maybe also toxic materials such as heavy metals. However, the total quantity of radioactive elements is less than in the original ore, and their collective radioactivity will be much shorter-lived.

Conversion and Enrichment

The uranium oxide product of a uranium mill is not directly usable as a fuel for a nuclear reactor and additional processing is required. Only 0.7% of natural uranium is 'fissile', or capable of undergoing fission, the process by which energy is produced in a nuclear reactor. The form, or isotope, of uranium which is fissile is the uranium-235 (U-235) isotope. The remainder is uranium-238 (U-238). For most kinds of reactor, the concentration of the fissile uranium-235 isotope needs to be increased – typically to between 3.5% and 5% U-235. Isotope separation is a physical process to concentrate ('enrich') one isotope relative to others. The enrichment process requires the uranium to be in a gaseous form. The uranium oxide concentrate is therefore first converted to uranium hexafluoride, which is a gas at relatively low temperatures. U-238 is fissionable in fast neutron reactors, which are likely to be in wide use by mid-century.

At a conversion facility, the uranium oxide is first refined to uranium dioxide, which can be used as the fuel for those types of reactors that do not require enriched uranium. Most is then converted into uranium hexafluoride, ready for the enrichment plant. The main hazard of this stage of the fuel cycle is the use of hydrogen fluoride. The uranium hexafluoride is then drained into 14-tonne cylinders where it solidifies. These strong metal containers are shipped to the enrichment plant.

The enrichment process separates gaseous uranium hexafluoride into two streams, one being enriched to the required level and known as low-enriched uranium; the other stream is progressively depleted in U-235 and is called 'tails', or simply depleted uranium.

The main enrichment process in commercial plants uses centrifuges, with thousands of rapidly-spinning vertical tubes. As they spin, the physical properties of molecules, specifically the 1% mass difference between the two uranium isotopes, separates them. A laser enrichment process is in the final stage of development.

The product of this stage of the nuclear fuel cycle is enriched uranium hexafluoride, which is reconverted to produce enriched uranium oxide. Up to this point the fuel material can be considered fungible (though enrichment levels vary), but fuel fabrication involves very specific design.

A small number of reactors, notably the Canadian and Indian heavy water type, do not require uranium to be enriched.

Fuel Fabrication

Reactor fuel is generally in the form of ceramic pellets. These are formed from pressed uranium oxide (UO_2) which is sintered (baked) at a high temperature (over 1400°C). The pellets are then encased in metal tubes to form fuel rods, which are arranged into a fuel assembly ready for introduction into a reactor. The dimensions of the fuel pellets and other components of the fuel assembly are precisely controlled to ensure consistency in the characteristics of the fuel.

In a fuel fabrication plant great care is taken with the size and shape of processing vessels to avoid criticality (a limited chain reaction releasing radiation). With low-enriched fuel criticality is most unlikely, but in plants handling special fuels for research reactors this is a vital consideration. Some 27 tonnes of fresh enriched fuel is required each year by a 1000 MWe reactor.

Power Generation and Burn-up

Several hundred fuel assemblies make up the core of a reactor. For a reactor with an output of 1000 megawatts (MWe), the core would contain about 75 tonnes of low-enriched uranium. In the reactor core the U-235 isotope fissions or splits, producing a lot of heat in a continuous process called a chain reaction. The process depends on the presence of a moderator such as water or graphite, and is fully controlled.

Some of the U-238 in the reactor core is turned into plutonium and about half of this is also fissioned, providing about one-third of the reactor's energy output (or more than half in CANDU).

As in fossil-fuel burning electricity generating plants, the heat is used to produce steam to drive a turbine and an electric generator, in a 1000 MWe unit providing over 8 billion kilowatt hours (8 TWh) of electricity in one year.

To maintain efficient reactor performance, about one-third of the spent fuel is removed every year or 18 months, to be replaced with fresh fuel. The length of fuel cycle is correlated with the use of burnable absorbers in the fuel, allowing higher burn-up.

In the USA about 85% of reactors have an 18-month fuel cycle, a few have 24-month ones. In Asia, over 80% have 18-month cycles, the rest 12-month. In Europe, over 60%

have 12-month cycles, the balance 18-month, and over one-quarter do not use burnable absorbers. All reactors in the USA and Asia use burnable absorbers. So 18 months is a typical worldwide refuelling interval.

Typically, some 44 million kilowatt-hours of electricity are produced from one tonne of natural uranium. The production of this amount of electrical power from fossil fuels would require the burning of over 20,000 tonnes of black coal or 8.5 million cubic metres of gas.

An issue in operating reactors and hence specifying the fuel for them is fuel burn-up. This is measured in gigawatt-days (thermal) per tonne and its potential is proportional to the level of enrichment. Hitherto a limiting factor has been the physical robustness of fuel assemblies, and hence burn-up levels of about 40 GWd/t have required only around 4% enrichment. But with better equipment and fuel assemblies, 55 GWd/t is possible (with 5% enrichment), and 70 GWd/t is in sight, though this would require 6% enrichment. The benefit of this is that operation cycles can be longer – around 24 months – and the number of fuel assemblies discharged as used fuel can be reduced by one third. Associated fuel cycle cost is expected to be reduced by about 20%.

In CANDU reactors using natural uranium, burn-up is much less, about 7.5 GWd/t, but in terms of efficiency this is equivalent to almost 50 GWd/t for enriched fuel.

Burn-up in GWd/t is the conventional measure for oxide fuels, and 60 GWd/t U is equivalent to about 6.5% atomic percent burn-up, *i.e.* about 6.5% of the original uranium atoms are burned directly, or indirectly via transformation to fissile plutonium. (With metal fuels, the atomic percent metric is used, and a new light water reactor metal fuel is targeting 21 atomic percent burn-up when it is deployed in 2020s.)

As with as a coal-fired power station about two thirds of the heat is dumped, either to a large volume of water (from the sea or large river, heating it a few degrees) or to a relatively smaller volume of water in cooling towers, using evaporative cooling (latent heat of vapourisation).

Used Fuel

With time, the concentration of fission fragments and heavy elements formed in the same way as plutonium in the fuel will increase to the point where it is no longer practical to continue to use the fuel, though much potential remains in it. So after 18-36 months the used fuel is removed from the reactor. The amount of energy that is produced from a fuel assembly varies with the type of reactor and the policy of the reactor operator. Used fuel will typically have about 1.0% U-235 and 0.6% fissile plutonium (almost 1% Pu total), with around 95% U-238. About 3% is fission products and minor actinides. Used fuel from PHWR units using natural (unenriched) uranium fuel after 7.5 GWd/t burn-up is 98.8% U-238, 0.23% U-235, 0.38% Pu and 0.6% fission products.

When removed from a reactor, the fuel will be emitting both radiation, principally from the fission fragments, and heat. It is unloaded into a storage pond immediately adjacent to the reactor to allow the radiation levels to decrease. In the ponds the water shields the radiation and absorbs the heat, which is removed by circulating the water to external heat exchangers. Used fuel is held in such pools for several months and sometimes many years. It may be transferred to naturally-ventilated dry storage on site after about five years.

Depending on policies in particular countries, some used fuel may be transferred to central storage facilities. Ultimately, used fuel must either be reprocessed in order to recycle most of it, or prepared for permanent disposal. The longer it is stored, the easier it is to handle, due to decay of radioactivity.

There are two alternatives for used fuel:

- Reprocessing to recover and recycle the usable portion of it.
- Long-term storage and final disposal without reprocessing.

Reprocessing

Used fuel still contains about 96% of its original uranium, of which the fissionable U-235 content has been reduced to less than 1%. About 3% of the used fuel comprises waste products and the remaining 1% is plutonium (Pu) produced while the fuel was in the reactor and not 'burned' then.

Reprocessing separates uranium and plutonium from waste products (and from the fuel assembly cladding) by chopping up the fuel rods and dissolving them in acid to separate the various materials. It enables recycling of the uranium and plutonium into fresh fuel, and produces a significantly reduced amount of waste (compared with treating all used fuel as waste). The remaining 3% of high-level radioactive wastes (some 750 kg per year from a 1000 MWe reactor) can be stored in liquid form and subsequently solidified.

Uranium and Plutonium Recycling

The uranium recovered from reprocessing, which typically contains a slightly higher concentration of U-235 than occurs in nature, can be reused as fuel after conversion and enrichment.

The plutonium can be directly made into mixed oxide (MOX) fuel, in which uranium and plutonium oxides are combined. In reactors that use MOX fuel, plutonium substitutes for the U-235 in normal uranium oxide fuel.

According to Areva, about eight fuel assemblies reprocessed can yield one MOX fuel assembly, two-thirds of an enriched uranium fuel assembly, and about three tonnes of depleted uranium (enrichment tails) plus about 150 kg of wastes. It avoids the need to purchase about 12 tonnes of natural uranium from a mine.

Another way of recycling plutonium and uranium from reprocessing is Russia's REMIX, not yet commercialized. Here a non-separated mix of both has some low-enriched uranium (17% U-235) added, to give fuel with about 1% Pu-239 and 4% U-235.

Apart from the incidental transformation of U-238 into plutonium in a normal reactor, U-238 is not usable in today's fuel cycle. However, in a fast neutron reactor it is fissionable as well as (more importantly) giving rise to plutonium, and is therefore potentially valuable. Increasingly, today's used fuel is being seen as a future resource rather than a waste.

Wastes

Wastes from the nuclear fuel cycle are categorised as high, medium or low-level wastes by the amount of radiation that they emit. These wastes come from a number of sources and include:

- Low-level waste produced at all stages of the fuel cycle;
- Intermediate-level waste produced during reactor operation and by reprocessing;
- High-level waste, which is waste containing the highly-radioactive fission products separated in reprocessing, and in many countries, the used fuel itself. Separated high-level wastes also contain long-lived transuranic elements.

After reprocessing, the liquid high-level waste can be calcined (heated strongly) to produce a dry powder which is incorporated into borosilicate (Pyrex) glass to immobilise it. The glass is then poured into stainless steel canisters, each holding 400 kg of glass. A year's waste from a 1000 MWe reactor is contained in five tonnes of such glass, or about 12 canisters 1.3 metres high and 0.4 metres in diameter. These can readily be transported and stored, with appropriate shielding.

The uranium enrichment process leads to the production of much 'depleted' uranium, in which the concentration of U-235 is significantly less than the 0.7% found in nature. Small quantities of this material, which is primarily U-238, are used in applications where high density material is required, including radiation shielding and some is used in the production of MOX fuel. While U-238 is not fissile it is a low specific activity radioactive material and some precautions must, therefore, be taken in its storage or disposal.

Used Fuel and Separated Wastes: Final Disposal

At the present time, there are no disposal facilities (as opposed to storage facilities) in operation in which used fuel, not destined for reprocessing, and the waste from reprocessing, can be placed. In either case the material is in a solid, stable wasteform.

Although technical issues related to disposal are straightforward, there is currently no pressing technical need to establish such facilities, as the total volume of such wastes is relatively small. Further, the longer it is stored the easier it is to handle, due to the progressive decrease of radioactivity.

There is also a reluctance to dispose of used fuel because it represents a significant energy resource which could be reprocessed at a later date to allow recycling of the uranium and plutonium.

A number of countries are carrying out studies to determine the optimum approach to the disposal of used fuel and wastes from reprocessing. The general consensus favours its placement into deep geological repositories, about 500 metres down, initially recoverable before being permanently sealed.

Other Sources of Nuclear Fuel

In the 1990s uranium mines gained a competitor, in many ways very welcome, as military uranium came on to the civil market under a US-Russian agreement. Since then half of the uranium used for electricity in the USA has come from Russian military stockpiles, and worldwide about one sixth of the market has been supplied thus.

Weapons-grade uranium in stockpiles built up during 1950s and 1960s has been enriched to more than 90% U-235 and must be diluted about 1:25 or 1:30 with depleted uranium (about 0.3% U-235). This means that progressively, Russian and other stockpiles of weapons material are used to produce electricity.

Weapons-grade plutonium may also be used to make mixed oxide (MOX) fuel for use in ordinary reactors or in special reactors designed to 'burn' it for electricity. Another US-Russian agreement covers disposition of military plutonium from both countries into MOX fuel.

Material Balance in the Nuclear Fuel Cycle

The following figures may be regarded as typical for the annual operation of a 1000 MWe nuclear power reactor typical of many operating today, using 4.5% enriched fuel and with 45 GWd/t burn-up:

Mining	Anything from 20,000 to 400,000 tonnes of uranium ore.
Milling	249 tonnes of uranium oxide concentrate (which contains 211 tonnes of uranium).
Conversion	312 tonnes of uranium hexafluoride, UF_6 (with 211 tU).
Enrichment	35.9 tonnes of enriched UF_6 (containing 24.3 t enriched U @ 4.5%) – balance is 'tails' @ 0.22%.
Fuel fabrication	27.6 tonnes UO_2 (with 24.3 t enriched U).

Reactor operation	8760 million kWh (8.76 TWh) of electricity at 100% output, hence 24 tonnes of natural U per TWh.
Used fuel	27.6 tonnes containing 280 kg transuranics (mainly plutonium), 26 t uranium oxide (<1.0% U-235), 1 tonne fission products.

The following figures assume the annual operation of 1000 MWe of nuclear power reactor capacity such as in the new AP1000 or EPR, with 5% enriched fuel and higher (65 GWd/t) burn-up:

Mining	Anything from 20,000 to 400,000 tonnes of uranium ore.
Milling	192 tonnes of uranium oxide concentrate (which contains 163 tonnes of uranium).
Conversion	241 tonnes of uranium hexafluoride, UF_6 (with 163 tU)
Enrichment	25 tonnes of enriched UF_6 (containing 16.85 t enriched U) – balance is 'tails' @ 0.22%.
Fuel fabrication	19.1 tonnes UO_2 (with 16.85 t enriched U).
Reactor operation	8760 million kWh (8.76 TWh) of electricity at 100% output, hence 18.6 tonnes of natural U per TWh.
Used fuel	19.1 tonnes containing 200 kg transuranics (mainly plutonium), 18.3 t uranium oxide (<1.0% U-235), <0.6 t fission products.

Between the above figures, efficiencies on power plant operation and lower enrichment tails assays meant that uranium demand per unit capacity was falling, and the report's generic reactor fuel consumption was reduced from 175 tU per GWe per year at 0.30% tails assay to 163 tU per GWe per year at 0.25% tails assay. The corresponding U_3O_8 figures are 206 tonnes and 192 tonnes per GWe per year.

Nuclear Fuel and its Fabrication

- Fuel fabrication is the last step in the process of turning uranium into nuclear fuel rods.
- Batched into assemblies, the fuel rods form the majority of a reactor core's structure.
- This transformation from a fungible material – uranium – to high-tech reactor components is conceptually different from the refining and preparation of fossil fuels.
- Nuclear fuel assemblies are specifically designed for particular types of reactors and are made to exacting standards.

- Utilities and fabricators have collaborated to greatly improve fuel assembly performance, and an international accident-tolerant fuel program is under way.
- While all present fuel is oxide, R&D is focused on metal, nitride and other forms. The first modern metal fuel is due to be trialled in commercial reactors.

Nuclear reactors are powered by fuel containing fissile material. The fission process releases large amounts of useful energy and for this reason the fissioning components – U-235 and/or Pu- 239 – must be held in a robust physical form capable of enduring high operating temperatures and an intense neutron radiation environment. Fuel structures need to maintain their shape and integrity over a period of several years within the reactor core, thereby preventing the leakage of fission products into the reactor coolant.

The standard fuel form comprises a column of ceramic pellets of uranium oxide, clad and sealed into zirconium alloy tubes. For light water reactor (LWR) fuel, the uranium is enriched to various levels up to about 4.8% U-235. Pressurised heavy water reactor (PHWR) fuel is usually unenriched natural uranium (0.7% U-235), although slightly-enriched uranium is also used.

Fuel assembly performance has improved since the 1970s to allow increased burn-up of fuel from 40 GWday/tU to more than 60 GWd/tU. This is correlated with increased enrichment levels from about 3.25% to 5% and the use of advanced burnable absorber designs for PWR, using gadolinium. Core monitoring giving detailed real-time information has enabled better fuel performance also.

The fabrication of fuel structures – called assemblies or bundles – is the last stage of the front end of the nuclear cycle shown in Figure, and represents less than 20% of the final cost of the fuel. The process for uranium-plutonium mixed oxide (MOX) fuel fabrication is essentially the same – notwithstanding some specific features associated with handling the plutonium component.

The industry is dominated by four companies serving international demand for light water reactors: Areva, Global Nuclear Fuel (GNF), TVEL and Westinghouse. GNF is mostly for BWR and TVEL for PWR.

There are three main stages in the fabrication of the nuclear fuel structures used in LWRs and PHWRs:

- Producing pure uranium dioxide (UO_2) from incoming UF_6 or UO_3.
- Producing high-density, accurately shaped ceramic UO_2 pellets.
- Producing the rigid metal framework for the fuel assembly – mainly from zirconium alloy; and loading the fuel pellets into the fuel rods, sealing them and assembling the rods into the final fuel assembly structure.

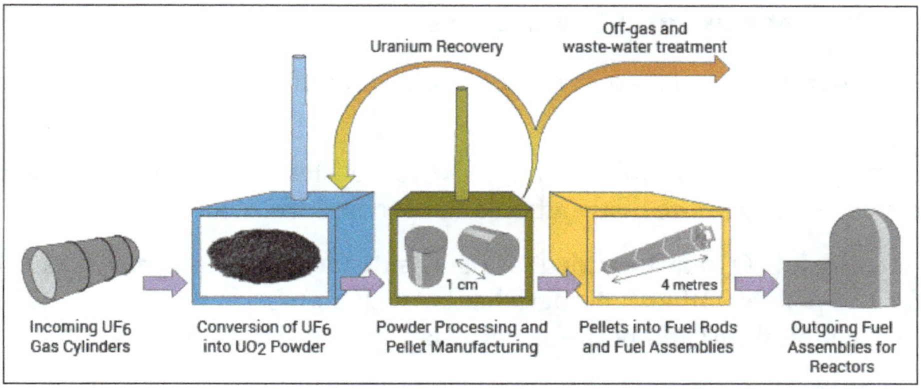

The fuel fabrication process.

UO_2 Powder Production

Uranium arrives at a fuel manufacturing plant in one of two forms, uranium hexafluoride (UF_6) or uranium trioxide (UO_3), depending on whether it has been enriched or not. It needs to be converted to uranium dioxide (UO_2) prior to pellet fabrication. Most fabrication plants have their own facilities for effecting this chemical conversion (some do not, and acquire UO_2 from plants with excess conversion capacity). Chemical conversion to and from UF_6 are distinct processes, but both involve the handling of aggressive fluorine compounds and plants may be set up to do both.

Conversion to UO_2 can be done using 'dry' or 'wet' processes. In the dry method, UF_6 is heated to a vapour and introduced into a two stage reaction vessel (eg, rotary kiln) where it is first mixed with steam to produce solid uranyl fluoride (UO_2F_2) – this powder moves through the vessel to be reacted with H_2 (diluted in steam) which removes the fluoride and chemically reduces the uranium to a pure microcrystalline UO_2 product.

Wet methods involve the injection of UF_6 into water to form a UO_2F_2 particulate slurry. Either ammonia (NH_3) or ammonium carbonate ($(NH_3)2CO_3$) is added to this mixture and the UO_2F_2 reacts to produce; ammonium diuranate (ADU, $(NH_3)2U_2O_7$) in the first case, or ammonium uranyl carbonate (AUC, $UO_2CO_3.(NH_3)2CO_3$) in the latter case. In both cases the slurry is filtered, dried and heated in a reducing atmosphere to pure UO_2. The morphology of UO_2 powders deriving from the ADU and AUC routes are different, and this has a bearing on final pellet microstructure.

Wet methods are slightly more complex and give rise to more wastes, however, the greater flexibility in terms of UO_2 powder properties is an advantage. For the conversion of UO_3 to UO_2, water is added to UO_3 so that it forms a hydrate. This solid is fed (wet or dry) into a kiln operating with a reducing atmosphere and UO_2 is produced.

Manufacture of Ceramic UO_2 Pellets

The UO_2 powder may need further processing or conditioning before it can be formed into pellets:

- Homogenization: Powders may need to be blended to ensure uniformity in terms of particle size distribution and specific surface area.
- Additives: U_3O_8 may be added to ensure satisfactory microstructure and density for the pellets. Other fuel ingredients, such as lubricants, burnable absorbers (e.g. gadolinium) and pore-formers may also need to be added.

Conditioned UO_2 powder is fed into dies and pressed biaxially into cylindrical pellet form using a load of several hundred MPa – this is done in pressing machines operating at high speed. These 'green' pellets are then sintered by heating in a furnace at about 1750 °C under a precisely controlled reducing atmosphere (usually argon-hydrogen) in order to consolidate them. This also has the effect of decreasing their volume. The pellets are then machined to exact dimensions – the scrap from which being fed back into an earlier stage of the process. Rigorous quality control is applied to ensure pellet integrity and precise dimensions.

For most reactors pellets are just under one centimetre in diameter and a little more than one centimetre long. A single pellet in a typical reactor yields about the same amount of energy as one tonne of steaming coal.

Burnable absorbers (or burnable 'poisons') such as gadolinium may be incorporated (as oxide) into the fuel pellets of some rods to limit reactivity early in the life of the fuel. Burnable absorbers have a very high neutron absorption cross-section and compete strongly for neutrons, after which they progressively 'burn-out' and convert into nuclides with low neutron absorption leaving fissile (U-235) to react with neutrons. Burnable absorbers enable longer fuel life by allowing higher fissile enrichment in fresh fuel, without excessive initial reactivity and heat being generated in the assembly.

Gadolinium, mostly at up to 3g oxide per kilogram of fuel, requires slightly higher fuel enrichment to compensate for it, and also after burn-up of about 17 GWd/t it retains about 4% of its absorbtive effect and does not decrease further. Zirconium diboride integral fuel burnable absorber (IFBA) as a thin coating on normal pellets burns away more steadily and completely, and has no impact on fuel pellet properties. It is now used in most US reactors and a few in Asia. China has this technology for AP1000 reactors.

Manufacture and Loading of the Fuel Assembly Framework

Nuclear fuel designs dictate that the pellet-filled rods have a precise physical arrangement in terms of their lattice pitch (spacing), and their relation to other features such as water (moderator) channels and control-rod channels. The physical structures for holding the fuel rods are therefore engineered with extremely tight tolerances. They must be resistant to chemical corrosion, high temperatures, large static loads, constant vibration, fluid and mechanical impacts. Yet they must also be as neutron-transparent as possible.

Assembly structures comprise a strong framework made from steel and zirconium upon which are fixed numerous grid support pieces that firmly hold rods in their precise lattice positions. These are made from zirconium alloy and must permit (and even enhance) the flow of coolant water around the fuel rod. The grid structures grip the fuel rod and so are carefully designed to minimise the risk of vibration-induced abrasion on the cladding tube – called 'fretting' wear. All fuel fabricators have highly sophisticated engineering processes and quality control for the timely manufacture of their assembly structures.

Pellets meeting QA specifications are loaded into tubes made from an appropriate zirconium alloy, referred to as the 'cladding'. The filled tube is flushed with helium and pressurized with tens of atmospheres (several MPa) of this gas before the ends are sealed at each end by precision welding. A free space is left between the top of the pellet stack and the welded end-plugs – this is called the 'plenum' space and it accommodates thermal expansion of the pellets and some fission product gases. A spring is usually put into the plenum to apply a compressive force on the pellet stack and prevent its movement. The completed fuel rods are then fixed into the prefabricated framework structures that hold the rods in a precisely defined grid arrangement.

In order to maximize the efficiency of the fission reaction the cladding and indeed all other structural parts of the assembly must be as transparent as possible to neutrons. Different forms of zirconium alloy, or zircaloy, are therefore the main materials used for cladding. This zircaloy includes small amounts of tin, niobium, iron, chromium and nickel to provide necessary strength and corrosion resistance. Hafnium, which typically occurs naturally with zirconium deposits, needs to be removed because of its high neutron absorption cross-section. The exact composition of the alloy used depends on the manufacturer and is an important determiner in the quality of the fuel assembly. Zircaloy oxidizes in air and water, and therefore it has an oxidized layer which does not impair function.

Safety Considerations

Rigorous quality control measures are employed at all process points in order to ensure traceability of all components in case of failures. The major process safety concerns at nuclear fuel fabrication facilities are those of fluoride handling and the risk of a criticality event if insufficient care is taken with the arrangement of fissile materials. Both risks are managed through the rigorous control of materials, indeed, fuel fabrication facilities operate with a strict limitation on the enrichment level of uranium that is handled in the plant – this cannot be higher than 5% U-235, essentially eliminating the possibility of inadvertent criticality.

Types of Nuclear Fuel Assemblies for Different Reactors

There is considerable variation among fuel assemblies designed for the different types of reactor. This means that utilities have limited choice in suppliers of fabricated fuel assemblies, especially for PWRs.

PWR Fuel

Pressurised water reactors (PWRs) are the most common type of nuclear reactor accounting for two-thirds of current installed nuclear generating capacity worldwide. A PWR core uses normal water as both moderator and primary coolant – this is kept under considerable pressure (about 10 MPa) to prevent it from boiling, and its temperature rises to about 330 °C after its upward passage past the fuel. It then goes through massive pipes to a steam generator.

Fuel for western PWRs is built with a square lattice arrangement and assemblies are characterized by the number of rods they contain, typically, 17×17 in current designs. A PWR fuel assembly stands between four and five metres high, is about 20 cm across and weighs about half a tonne. The assembly has vacant rod positions – space left for the vertical insertion of a control rod. Not every assembly position requires fuel or a control rod, and a space may be designated as a "guide thimble" into which a neutron source rod, specific instrumentation, or a test fuel segment can be placed.

A PWR fuel assembly comprises a bottom nozzle into which rods are fixed through the lattice and to finish the whole assembly it is crowned by a top nozzle. The bottom and top nozzles are heavily constructed as they provide much of the mechanical support for the fuel assembly structure. In the finished assembly most rod components will be fuel rods, but some will be guide thimbles, and one or more are likely to be dedicated to instrumentation. A PWR fuel assembly is shown in Figure. PWR fuel assemblies are rather uniform compared with BWR ones, and those in any particular reactor must have substantially the same design.

An 1100 MWe PWR core may contain 193 fuel assemblies composed of over 50,000 fuel rods and some 18 million fuel pellets. Once loaded, fuel stays in the core for several years depending on the design of the operating cycle. During refueling, every 12 to 18 months, some of the fuel - usually one third or one quarter of the core – is removed to storage, while the remainder is rearranged to a location in the core better suited to its remaining level of enrichment.

Russian PWR reactors are usually known by the Russian acronym VVER. Fuel assemblies for these are characterized by their hexagonal arrangement, but are otherwise of similar length and structure to other PWR fuel assemblies. Most is made by TVEL in Russia, but Westinghouse in Sweden also fabricates it and is increasing capacity to do so. TVEL is instigating using erbium as a burnable poison in fuel enriched to about 6.5% in order to prolong the intervals between refuelling to two years.

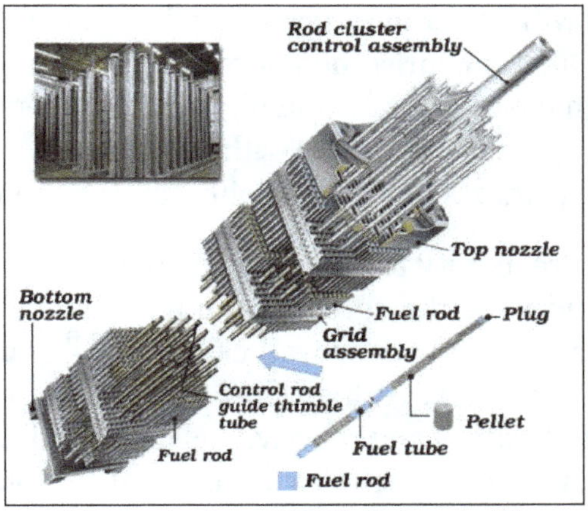

Schematic view of PWR fuel assembly (Mitsubishi Nuclear Fuel).

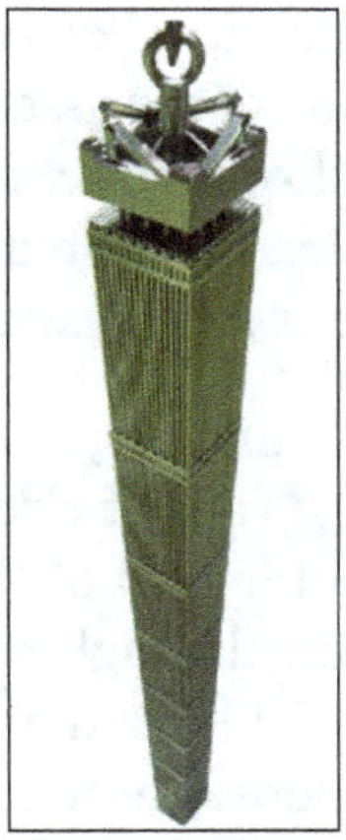

A PWR fuel assembly.

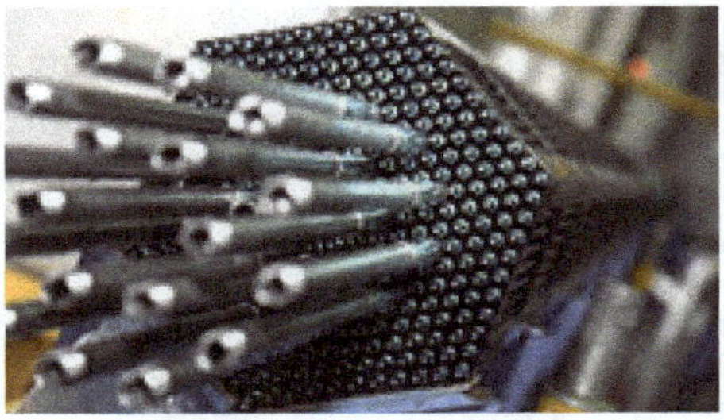

A VVER-1000 fuel assembly.

BWR Fuel

Boiling water reactors (BWRs) are the second most common nuclear reactor type accounting for almost one-quarter of installed nuclear generating capacity. In a boiling water reactor water is turned directly to steam in the reactor pressure vessel at the top of the core and this steam (at about 290 °C and 7 MPa) is then used to drive a turbine.

BWRs also use fuel rods comprising zirconium-clad uranium oxide ceramic pellets. Their arrangement into assemblies is again based on a square lattice, with pin geometries ranging from 6x6 to 10x10 or 11x11. Fuel life and management strategy is similar to that for a PWR.

But BWR fuel is fundamentally different from PWR fuel in certain ways: (i) Four fuel assemblies and a cruciform shaped control blade form a 'fuel module', (ii) Each assembly is isolated from its neighbours by a water-filled zone in which the cruciform control rod blades travel (they are inserted from the bottom of the reactor), (iii) Each BWR fuel assembly is enclosed in a zircaloy sheath or channel box which directs the flow of coolant water through the assembly and during this passage it reaches boiling point, (iv) BWR assemblies contain larger diameter water channels – flexibly designed to provide appropriate neutron moderation in the assembly.

The zircaloy tubes are allowed to fill with water thus increasing the amount of moderator in the central region of the assembly. Different enrichment levels are used in the rods in varying positions – lower enrichments in the outer rods, and higher enrichments near the centre of the bundle. A BWR reactor is designed to operate with 12-15% of the water in the top part of the core as steam, and hence with less moderating effect and thus efficiency there.

For many BWR models, control of reactivity to enable load-following can be achieved by changing the rate of circulation inside the core. Jet pumps located in the annulus between the outer wall of the vessel and an inner wall called the shroud increase the flow of water up through the fuel assembly. At high flow rates steam bubbles are removed more quickly, and hence moderation and reactivity is increased. When flow rate is decreased, moderation decreases as steam bubbles are present for longer and hence reactivity drops. This allows for a variation of about 25% from the maximum rated power output, enabling load-following more readily than with a PWR.

Control rods are used when power levels are reduced below 75%, but they are not part of the fuel assembly as in a PWR. They are bottom-entry – pushed upwards so that rods intercept the lower, more reactive, zone of the fuel assemblies first. BWR fuel fabrication takes place in much the same way as PWR fuel.

BWR fuel assemblies therefore operate more as individual units, and different designs may be mixed in any core load, giving more flexibility to the utility in fuel purchases.

GE's Global Nuclear Fuels is developing fuel with new clad material – NSF – containing 1% niobium, 1% tin, 0.35% iron (Nb,Sn,Fe) to reduce or eliminate fuel channel distortion due to chemical interaction with zircaloy and in 2013, 8% of cores were using this. Toshiba and ceramics company Ibiden in Japan are developing silicon carbide sheaths or channel boxes for BWR fuel assemblies.

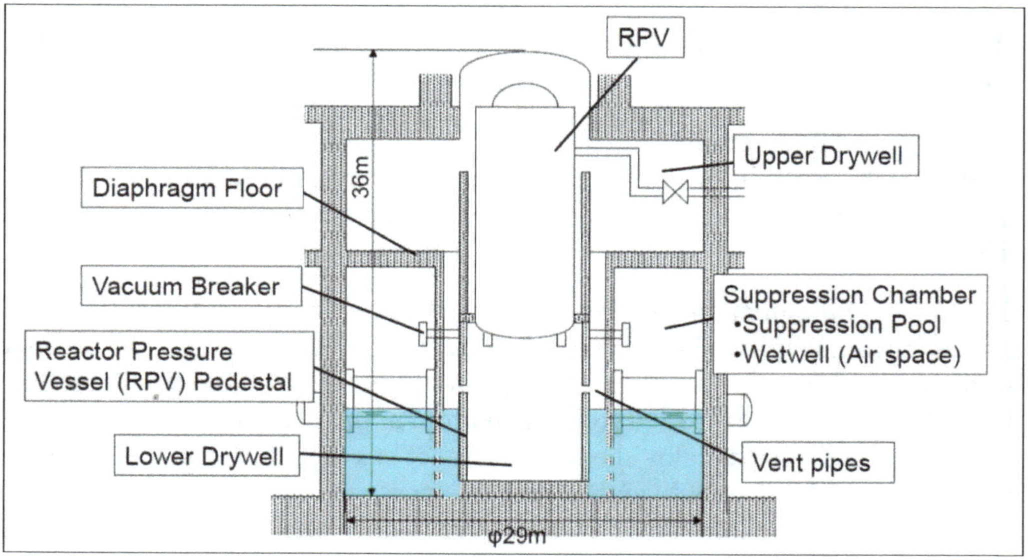

Schematic view of BWR fuel assembly (Nucleartourist and GE).

This has a low-tin zirconium channel material and new fuel cladding. It says that this fuel features improved economy, robust mechanical design and high-performing material. It has been optimised for both short- and long-cycle operation as well as for uprated cores and higher burn-ups.

PHWR (CANDU) Fuel

Pressurised Heavy Water Reactors (PHWRs) are originally a Canadian design (also called "CANDU") accounting for ~6% of world installed nuclear generating capacity. PHWRs use pressure tubes in which heavy water moderates and cools the fuel. They run on natural (unenriched) or slightly-enriched uranium oxide fuel in ceramic pellet form, clad with zirconium alloy.

PHWR fuel rods are about 50 cm long and are assembled into 'bundles' approximately 10 cm in diameter. A fuel bundle comprises 28, 37 or 43 fuel elements arranged in several rings around a central axis. Their short length means that they do not require the support structures characteristic of other reactor fuel types. PHWR fuel does not attain high burn-up, nor does it reside in the reactor core for very long and so the fuel pellets swell very little during their life. This means that PHWR fuel rods do not need to maintain a pellet-cladding gap, nor be highly pressurized with a filling gas (as for LWR fuel), indeed, the metal cladding is allowed to collapse onto the fuel pellet thereby assuring good thermal contact.

The fuel bundles are loaded into horizontal channels or pressure tubes which penetrate the length of the reactor vessel (known as the calandria), and this can be done while the reactor is operating at full power. About twelve bundles are loaded into each fuel channel depending on the model – a 790 MWe CANDU reactor contains 480 fuel channels composed of 5,760 fuel bundles containing over 5 million fuel pellets.

The on-load refueling is a fully-automated process: new fuel is inserted into a channel at one end and used fuel is collected at the other. This feature means that the PHWR is inherently flexible with its fuel requirements, and can run on different fuels requiring different residence times, eg natural uranium, slightly enriched uranium, plutonium-bearing fuels and thorium-based fuels.

Indian PHWR fuel bundles.

AGR Fuel

The Advanced Gas-Cooled Reactor (AGR) is a second-generation UK-designed nuclear reactor only used in UK. AGRs account for about 2.7% of total global nuclear generating capacity. They employ a vertical fuel channel design, and use CO_2 gas – a very weak moderator – as the primary coolant.

AGR fuel assemblies consist of a circular array of 36 stainless steel clad fuel pins each containing 20 enriched UO_2 fuel pellets, and the assembly weighs about 43 kilograms. Enrichment levels vary up to about 3.5%. Stainless steel allows for higher operating temperatures but sacrifices some neutron economy. The assembly is covered with a graphite sheath which serves as a moderator. Eight assemblies are stacked end on end in a fuel channel, inserted down through the top of the reactor. During refueling this whole stack is replaced. Fuel life is about five years, and refueling can be carried out on-load through a refuelling machine.

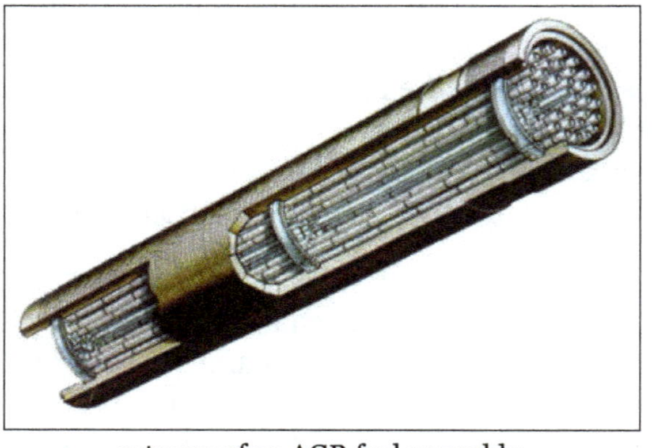

cutaway of an AGR fuel assembly.

RBMK Fuel

The RBMK reactor is an early Soviet design, developed from plutonium production reactors. Eleven units are in operation (3% of world total), with control systems and oxide fuel greatly modified since 1990. It employs vertical pressure tubes (just under 1700 of these, each about 7 metres long) running through a large graphite moderator. The fuel is cooled by light water water, which is allowed to boil in the primary circuit, much as in a BWR.

RBMK fuel rods are about 3.65 metres long, and a set of 18 forms a fuel bundle about 8 cm diameter. Two bundles are joined together and capped at either end by a top and bottom nozzle, to form a fuel assembly with an overall length of about 10 metres, weighing 185 kilograms. Since 1990 RBMK fuel has had a higher enrichment level, increasing from about 2% to average 2.8% (varying along the fuel element from 2.5% to 3.2%) and it now includes about 0.6% erbium (a burnable absorber). This has the effect of improving overall safety and increasing fuel burn-up. This new fuel can stay in the reactor for periods of up to six years before needing to be removed. All RBMK reactors now use recycled uranium from VVER reactors. As with other pressure tube designs such as the PHWR, the RBMK reactor is capable of on-load refueling.

Fast Neutron Reactor Fuel

There is only one commercially operating fast reactor (FNRs) in service today – the BN-600 at Beloyarsk in Russia. There are two FNRs under construction – a 800 MWe unit in Russia and a 500 MWe unit in India (which expects to build five more). Two BN-800 units were planned in China.

Fast neutron reactors (FNR) are unmoderated and use fast neutrons to cause fission. Hence they mostly use plutonium as their basic fuel, or sometimes high-enriched uranium to start them off (they need about 20-30% fissile nuclei in the fuel). The plutonium is bred from U-238 during operation. If the FNR is configured to have a conversion ratio above 1 (ie more fissile nuclei are created than fissioned) as originally designed, it is called a Fast Breeder Reactor (FBR). FNRs use liquid metal coolants such as sodium and operate at higher temperatures.

Apart from the main FNR fuel, there are numerous heavy nuclides - notably U-238, but also Am, Np and Cm that are fissionable in the fast neutron spectrum – compared with the small number of fissile nuclides in a slow (thermal) neutron field (just U-235, Pu-239 and U-233). A FNR fuel can therefore include a mixture of transuranic elements. Also it can be in one of several chemical forms, including; standard oxide ceramic, mixed oxide ceramic (MOX), single or mixed nitride ceramics, carbides and metallic fuels. Further, FNR fuel can be fabricated in pellet form or using the 'vibro-pack' method in which graded powders are loaded and compressed directly into the cladding tube. Carbide fuels such as used in India have a higher thermal conductivity than oxide fuels and can attain breeding ratios larger than those of oxide fuels but less than metal fuels.

The core of an FNR is much smaller than a conventional reactor, and cores tend to be designed with distinct 'seed' and 'blanket' regions according to whether the reactor is to be operated as a 'burner' or a 'breeder'. In each case the fuel composition for the seed and blanket regions are different – the central seed region uses fuel with a high fissile content (and thus high power and neutron emission level), and the blanket region has a low fissile content but a high level of neutron absorbing material which can be fertile (for a breeding design, eg U-238) or a waste absorber to be transmuted.

BN-600 fuel assemblies are 3.5 m long, 96 mm wide, weigh 103 kg and comprise top and bottom nozzles (to guide coolant flow) and a central fuel bundle. The central bundle is a hexagonal tube and for seed fuel houses 127 rods, each 2.4 m long and 7 mm diameter with ceramic pellets in three uranium enrichment levels; 17%, 21% and 26%. Blanket fuel bundles have 37 rods containing depleted uranium. BN-600 fuel rods use low-swelling stainless steel cladding.

FNRs use liquid metal coolants such as sodium or a lead-bismuth eutectic mixture and these allow for higher operating temperatures – about 550 °C, and thus have higher energy conversion efficiency. They are capable of high fuel burn-up.

Nuclear Fuel Performance in Reactors

Nuclear fuel operates in a harsh environment in which high temperature, chemical corrosion, radiation damage and physical stresses may attack the integrity of a fuel assembly. The life of a fuel assembly in the reactor core is therefore regulated to a burn-up level at which the risk of its failure is still low. Fuel 'failure' refers to a situation when the cladding has been breached and radioactive material leaks from the fuel ceramic (pellet) into the reactor coolant water. The radioactive materials with most tendency to leak through a cladding breach into the reactor coolant are fission-product gases and volatile elements, notably; krypton, xenon, iodine and cesium.

Fuel leaks do not present a significant risk to plant safety, though they have a big impact on reactor operations and (potentially) on plant economics. For this reason, primary coolant water is monitored continuously for these species so that any leak is quickly detected. The permissible level of released radioactivity is strictly regulated against specifications which take into account the continuing safe operation of the fuel. Depending on its severity a leak will require different levels of operator intervention:

1. Very minor leak: No change to operations – the faulty fuel assembly with leaking rod(s) is removed at next refuelling, inspected, and possibly re-loaded.
2. Small leak: Allowable thermal transients for the reactor are restricted. This might prevent the reactors from being able to operate in a 'load-follow' mode and require careful monitoring of reactor physics. The faulty fuel assembly with leaking rod is generally removed and evaluated at the next scheduled refuelling.
3. Significant leak: The reactor is shut down and the faulty assembly located and removed.

A leaking fuel rod can sometimes be repaired but it is more usual that a replacement assembly is needed (this having a matching level of remaining enrichment). Replacement fuel is one cost component associated with failed fuel. There is also the cost penalty and/or replacement power from having to operate at reduced power or having an unscheduled shutdown. There may also be higher operation and maintenance costs associated with mitigating increased radiation levels in the plant.

Fuel management is a balance between the economic imperative to burn fuel for longer and the need to keep well within failure-risk limits. Improving fuel reliability extends these limits, and therefore is a critical factor in providing margin to improve fuel burn-up.

The nuclear industry has made significant performance improvements reducing fuel failure rates by about 60% in the 20 years to 2006 to an average of some 14 leaks per million rods loaded. The reliability drive continues. Industry-wide programs led by the Electric Power Research Institute (EPRI) and the US Institute of Nuclear Power Operations (INPO) have produced guidelines to help eliminate fuel failures (there was an ambitious goal to achieve zero fuel failures). These programs led to the accident-tolerant fuel program.

Fuel failures in US power reactors are rare. As of early 2014, 97% of US nuclear power plants were free of fuel failures, compared with 71% in 2007, according to EPRI. The annual US failure rate is about one in one million (*i.e.* five rods per year). Fuel engineering continues to improve, *e.g.* with more sophisticated debris filters in assembly structures. Utilities themselves impose more rigorous practices to exclude foreign material entering primary coolant water. Global Nuclear Fuel (GNF) in 2013 had two million fuel rods in operation and claimed to have no leakers among them. (In the early 1970s hydriding and pellet-clad interaction caused a lot of leaks. The 1980s saw an order of magnitude improvement.)

At the same time there has been a gradual global trend toward higher fuel burnup. However, higher burnup generally requires higher enrichment levels, and there is a limit on this given the strict criticality safety limitation imposed on fuel fabrication facilities – the maximum uranium enrichment level that can be handled is normally 5% U-235.

An equally important trend in nuclear fuel engineering is to be able to increase the power rating for fuels, ie, how much energy can be extracted per length of fuel rod. Currently this is limited by the material properties of the zirconium cladding.

Fabrication Supply and Demand

The current annual demand for LWR fuel fabrication services is expressed as a requirement for about 7000 tonnes of enriched uranium being made into assemblies, and this is expected to increase to about 9500 t by 2020. Requirements for PHWRs account for an additional 3,000 t/year and the Gas-cooled Reactor market for around 400 t/year.

Requirements for fuel fabrication services will grow roughly in line with the growth in nuclear generating capacity. However, fabrication requirements are also affected by changes in utilities' reactor operating and fuel management strategies, which are partly driven by technical improvements in fuel fabrication itself. For example, LWR discharge burn-ups have increased steadily as improvements in fuel design have made this possible, and this has tended to reduce fabrication demand, as fuel remains in the reactor for a longer period (though there is a limit to how far burn-ups can be pushed without tackling the 5% enrichment limit in place for criticality safety margins at fuel fabrication plants). A parallel industry-wide focus on increasing fuel performance and reliability has also decreased the demand for fuel to replace defective assemblies.

Plans to build many new reactors affect the demand on fabrication capacity in two ways. The demand for reloads increases in line with the new installed reactor capacity, typically 16 to 20 tonnes per year per GW. Additionally the first cores create a temporary peak demand, since the amount required is about three to four times that of a reload batch in currently operated LWRs (and some of the enrichment is less). An average first core enrichment is about 2.8%.

Secondary Supply from Recycle

Currently about 100 t/yr year of reprocessed uranium (RepU) is produced at MSZ in Elektrostal, Russia (capacity 250 t/yr) for AREVA contracts. One production line in AREVA's plant in Romans, France is licensed to fabricate 150 t of RepU into fuel per year and PWR assemblies of this type have already been delivered to French, Belgian and UK reactors, and an amount of RepU powder has been sent from Russia to Japan. Limited RepU and enriched RepU (ERU) capacity exists elsewhere also.

At present, nearly all commercial MOX fuel is fabricated in AREVA's MELOX plant in Marcoule. With a capacity of 195 tonnes/yr and a good production rate this plant helps not only to save uranium and enrichment demand, but also frees up LWR fabrication capacity in the market.

The UK's Sellafield MOX plant had a designed capacity of 120 t/yr but was downgraded to 40 t/yr and never reached that level of reliable output before being closed down in 2011. Russia's MOX plant at Zheleznogorsk for fast reactors commenced operation in 2015. Japan's Rokkasho-Mura MOX plant is planned to be and the US MOX plant in Savannah River was due to produce MOX fuel from weapons plutonium but this project has now been terminated.

The MOX fuel market has weakened somewhat recently with the cessation of its use in Belgium, Germany and Switzerland (moratorium), and the continued loading of MOX fuel in Japan has diminished in the aftermath of the Fukushima accident.

Table: World MOX fuel fabrication capacity, tonnes/yr.

	Fabricator	Location	Pelletising	Rod/assembly
France	Areva NC	Marcoule	195	195
India	DAE Nuclear Fuel Complex	Tarapur	50	50
Japan	JAEA	Tokai-Mura	5	5
	JNFL	Rokkasho-Mura	130	130
Russia	MCC	Zheleznogorsk	60	60
Total			440	440

MOX Fuel

Mixed uranium oxide + plutonium oxide (MOX) fuel has been used in about 30 light-water power reactors in Europe and about ten in Japan. It consists of depleted uranium (about 0.2% U-235), large amounts of which are left over from the enrichment of uranium, and plutonium oxide that derives from the chemical processing of used nuclear fuel (at a reprocessing plant). This plutonium is reactor-grade, comprising about one third non-fissile isotopes.

In a MOX fuel fabrication plant the two components are vigorously blended in a high-energy mill which intimately mixes them such that the powder becomes mainly a single 'solid solution' $(U,Pu)O_2$. MOX fuel with about 7% of rector-grade plutonium is equivalent to a typical enriched uranium fuel. The pressing and sintering process is much the same as for UO_2 fuel pellets, though some plastic shielding is needed to protect workers from spontaneous neutron emissions from the Pu-240 component.

Vibropacked MOX (VMOX) fuel is a Russian variant for MOX fuel production in which blended $(U,Pu)O_2$ and UO_2 powders are directly loaded and packed into cladding tubes where they sinter in-situ under their own operating temperature. This eliminates the need to manufacture pellets to high geometric tolerances, which involves grinding and scrap which are more complex to deal with for Pu-bearing fuels. Russian sources say vibropacked fuel is more readily recycled.

REMIX Fuel

REMIX (from Regenerated Mixture) fuel is produced directly from a non-separated mix of recycled uranium and plutonium from reprocessing used fuel, with a LEU (up to 17% U-235) uranium make-up comprising about 20% of the mix. This gives fuel with about 1% Pu-239 and 4% U-235 which can sustain burn-up of 50 GWd/t over four years. The spent REMIX fuel after four years is about 2% Pu-239 and 1% U-235, and following cooling and reprocessing the non-separated uranium and plutonium is recycled again after LEU addition, which compensates for the even isotopes of both elements.

REMIX-fuel can be repeatedly recycled with 100% core load in current VVER-1000 reactors and, correspondingly reprocessed many times – up to five times, so that with three fuel loads in circulation a reactor could run for 60 years using the same fuel, with LEU recharge. REMIX can serve as a replacement for existing reactor fuel, though it is not yet commercialised. The REMIX cycle can be modified from the above figures according to need.

TRISO High Temperature Reactor Fuel

High temperature reactors (HTR) operate at 750 to 950 °C, and are normally helium-cooled. Fuel for these is in the form of TRISO (tristructural-isotropic) particles less than a millimetre in diameter. Each has a kernel (ca. 0.5 mm) of uranium oxycarbide (or uranium dioxide), with the uranium enriched up to 20% U-235, though normally less. This is surrounded by layers of carbon and silicon carbide, giving a containment for fission products which is stable to very high temperatures. Recent trials at two US laboratories confirmed that most fission products remain securely in TRISO particles up to about 1800 °C – the performance being much better than previously known.

There are two ways in which these particles can be arranged in a HTR: in blocks – hexagonal 'prisms' of graphite, or in billiard ball-sized pebbles of graphite encased in silicon carbide, each with about 15,000 fuel particles and 9g uranium. Either way, the moderator is graphite. HTRs can potentially use thorium-based fuels, such as high-enriched or low-enriched uranium with Th, U-233 with Th, and Pu with Th. Most of the experience with thorium fuels has been in HTRs.

The main HTR fuel fabrication plant is at Baotou in China, the Northern Branch of China Nuclear Fuel Element Co Ltd. From 2015 this makes 300,000 fuel pebbles per year for the HTR-PM under construction at Shidaowan. Previous production has been on a small scale in Germany. In the USA, BWX Technologies is making TRISO fuel on an engineering scale, funded by the DOE which is aiming to take it to commercial scale. In September 2017 X-energy signed an agreement with Centrus Energy to develop TRISO fabrication technology for uranium carbide fuel. In Japan, NFI at Tokai has 400 kgU/yr HTR fuel capacity.

Other High-assay Low-enriched Fuel

In connection with a number of small modular reactor (SMR) designs, attention is turning to the need for high-assay low-enriched uranium (HALEU), with enrichment levels between 5% and 20% U-235. In the USA, the Department of Energy (DOE) is proposing to convert metallic HALEU into fuel for research and development purposes at Idaho National Laboratory's Materials and Fuels Complex and/or the Idaho Nuclear Technology and Engineering Center, to support the development of new reactor technologies with higher efficiencies and longer core lifetimes. HALEU may be metallic or oxide.

HALEU can be produced with existing centrifuge technology, but a number of arrangements would need to be made for this, as well as for deconversion and fuel fabrication. New transport containers would also be required as those for today's enriched UF_6 could not be used due to criticality considerations.

Advanced Nuclear Fuel Technology

Fuel development activities in the nuclear industry have largely focused on improving the reliability of standard zirconium-clad uranium oxide fuels. Increasingly, however, R&D effort is being applied to evolutionary fuel forms that can offer significant improvements in terms of safety, waste management and operating economics, as well as allowing the deployment of new types of reactor.

Accident Tolerant Fuel

Accident tolerant fuel (ATF) is a term used to describe new technologies that enhance the safety and performance of nuclear fuel. ATF may incorporate the use of new materials and designs for cladding and fuel pellets.

Framatome, GE/GNF and Westinghouse are all developing ATF concepts with the help of funding from the US Department of Energy (DOE). Since 2012, the DOE has supported the development of ATF concepts through its Enhanced Accident Tolerant Fuel (EATF) programme. Its objective is to develop new cladding and fuel materials that can better tolerate the loss of active cooling in the core, while maintaining or improving fuel performance and economics during normal operations. A priority of the EATF programme is to minimise the generation of hydrogen.

The joint DOE programme uses the Halden research reactor in Norway to test ATF fuel rods, as well as the Advanced Test Reactor (ATR) and the restarted Transient Reactor Test Facility (TREAT) at the DOE's Idaho National Laboratory (INL). In February 2017 the DOE awarded $10 million to Framatome (then Areva) over two years for phase 2 of the programme, and similar funding is being provided for GE Hitachi and Westinghouse.

Framatome in phase 2 of the ATF programme from 2017 has been developing a nuclear fuel concept, using a chromium-coated zirconium alloy cladding combined with chromia-doped fuel pellets. The fuel is expected to retain fission gases better and improve pellet-cladding interaction, and the cladding will better resist high-temperature oxidation. In June 2018 the DOE announced testing of Framatome's ATF in the Advanced Test Reactor at Idaho National Laboratory. The first full Gaia test assemblies were loaded into Southern Nuclear's Vogtle 2 in March 2019. Framatome is also continuing work on a silicon carbide cladding, and plans to use that cladding on chromia-doped pellets in lead test assemblies in about 2022.

GE Hitachi with GNF is developing a ferritic/martensitic steel alloy cladding (*e.g.* Fe-Cr-Al) for conventional UO_2 fuel. The Fe-Cr-Al cladding has better mechanical strength at high temperatures, retains fission gases better than zirconium alloy and has less potential for hydrogen generation in an accident. GNF's advanced fuel rods were shipped to Southern Nuclear at the start of 2018. They were the first developed through the ATF programme to be loaded into a commercial reactor during Hatch 1's spring refuelling outage in March 2018.

Westinghouse in June 2017 launched its EnCore ATF. It is manufacturing the first En-Core test rods, with lead test assembly insertion of these to start in 2019, at Exelon's Byron plant. The initial EnCore fuel comprises high-density uranium silicide fuel pellets inside zirconium cladding with a thin coating of chromium making it more robust chemically. (Uranium silicide – U_3Si_2 – fuels for research reactors are being developed at INL also.) In the second phase, the uranium silicide fuel pellets would be in silicon carbide ceramic matrix composite cladding with a melting point of 2800 °C, and these test assemblies could be loaded into a reactor by 2022. The EnCore ATF fuel development has been supported by awards from the DOE to Westinghouse and a group of partners including General Atomics, several DOE national laboratories, Southern Nuclear Operating Company and Exelon.

Westinghouse said that cost savings will arise because the uranium silicide offers up to 20% higher density of uranium and much higher thermal conductivity which does not degrade with irradiation like UO_2, so fewer fuel assemblies need to be replaced during each refuelling outage. In the second phase of EnCore, the higher temperature tolerance of silicon carbide cladding has potential for revised regulatory requirements, and Westinghouse sees this as a "game changer".

After trials of lead test assemblies, Westinghouse intends to make full reload quantities available from 2027. ATFs present a number of manufacturing challenges and, given the neutronic penalties entailed, enrichment to over 5% could be needed, despite the higher density of uranium in the fuel.

Rosatom's fuel company TVEL plans to offer ATF to its customers by the early 2020s. TVEL is developing ATF for use in Rosatom's VVER reactors and in Western PWRs. Prototype assemblies – one VVER and one for western PWR – are being tested at the MIR research reactor at the Research Institute of Atomic Reactors at Dimitrovgrad. TVEL expects to finalise testing late in 2019, at which point it would decide on the technology to be used for its commercial offering. TVEL stated that its ATF fuel has a heat-resistant coating on the zirconium cladding and fuel pellet composition may include molybdenum or uranium disilicide. Is likely to involve a change to both cladding material and fuel pellet composition.

The OECD Nuclear Energy Agency's Expert Group on ATFs for Light Water Reactors reviews cladding and core materials focusing on their fundamental properties and behaviour under normal operation and accident conditions, as described above for the US-led program. Both advanced core materials and components, in particular innovative cladding materials (coated and improved Zrbased alloys, SiC and SiC/SiC composites, advanced steels and refractory metals) and nonfuel components (advanced control rods, BWR channel box) are considered. A sub-group focuses on fuel design to address three categories of innovative fuels: Improved UO_2, highdensity fuels and coatedparticle fuels such as the HTR fuel.

Metal Fuel

Independently of the US DOE and the international ATF program, Lightbridge is developing an advanced metal fuel concept that may have accident tolerant characteristics.

Metal fuels were used in some earlier reactors such as the UK Magnox design, and also in two US fast reactors, with 5-10% zirconium alloyed. But the higher melting point of uranium oxide has made it the preferred fuel in all reactors for half a century. At least in the USA, metal fuels have not been made since the 1980s.

However, metal has much better thermal conductivity than ceramic oxide, and recent research has turned back to metal fuel forms. Babcock & Wilcox Nuclear Energy was working with Lightbridge to set up a pilot plant for metal fuel which is 50:50 (by mass) Zr-U alloy, with uranium enriched to almost 20% and having a multi-lobed and helically twisted rod geometry. The increased enrichment compensates for reductions in the initial fissile loading and in the derivative plutonium. Melting point of the alloy is about 1600 °C, and average operating temperature in the fuel is up to 370 °C (rather than about 1250 °C in normal oxide fuel), the thermal conductivity being five times better than oxide fuel. BWXT in the USA has now completed its assessment of feasibility and prepared a fabrication plan for manufacturing fuel samples.

Each Lightbridge fuel rod consists of a central displacer of zirconium surrounded by a four-lobed fuel core with the cladding metallurgically bonded to it. For the hexagonal fuel assemblies for VVERs, the fuel rod is three-lobed. The shape of the rod provides increased surface area for heat transfer and the area between the lobes accommodated swelling during irradiation. The rod has greater structural integrity than current tubes with ceramic pellets inside. The twist of about 180° over about a metre means that the rods are self-spacing while giving good flow characteristics. The fuel operates at a higher power density than oxide fuels and the target burn-up is 21 atomic percent, about three times that of oxide fuels. It is suitable for all LWRs, and is expected to give a power uprate of about 17% in existing PWRs, and up to 30% in new ones designed for the higher power density, with longer fuel cycle. In adition to US and Russian patents, in June 2015 the design was patented in South Korea, where Lightbridge sees a "significant potential market", and in July 2017 this patent protection was extended to cover both the metallic four-lobe design and its manufacture from powder. By November 2017 it had been patented in Japan, China (four patents), South Korea, and Canada.

Lightbridge has also agreed with Canadian Nuclear Laboratories for fabrication of such metal fuel at Chalk River in Canada and testing of it there in the NRU reactor. The agreement was expected to see fabrication and characterization of prototype fuel rods using depleted uranium in early 2016, with irradiation fuel samples using enriched uranium made later the same year. Subject to final approval from the Norwegian Radiation Protection Authority, Lightbridge will test the fuel under prototypical PWR conditions in a pressurized water loop of Norway's 25 MWt Halden research reactor (a boiling heavy water reactor). The initial phase of irradiation testing was to begin in 2017 using short samples to evaluate conductivity, and continue for about three years using 70 cm fuel rods to evaluate cladding and swelling. Tests aim to reach the burnup necessary for insertion of lead test assemblies (LTAs) in a commercial power reactor. The final phase of irradiation testing necessary for batch reloads and full cores operating with a 10% power uprate and a 24-month cycle is expected to take an additional two years and be completed while LTAs have begun operating in the core of a commercial power reactor about 2020.

In April 2015 a group of electric utilities representing half of the US nuclear generating capacity wrote to the Nuclear Regulatory Commission formally expressing interest in Lightbridge's metal fuel, saying that they believed the fuel provided opportunities to improve safety and fuel cycle economics significantly. The utilities' Nuclear Utility Fuel Advisory Board (NUFAB) expects to test the fuel in an operating PWR about 2020.

In March 2016 Lightbridge entered into an exclusive joint development agreement with Areva NP to set up a 50:50 joint venture that would develop, fabricate and commercialize fuel assemblies based on the metallic fuel technology. In November it announced an agreement on key terms for the US-based joint venture, creating "a viable and well-defined commercialization path" covering fuel assemblies for most types of light water reactors, including pressurized water reactors (excluding VVERs), boiling water reactors, small modular reactors and research reactors. In September 2017 a binding agreement was signed with Areva Inc (for New NP) to set up the joint venture in North America. The joint venture between Lightbridge and Framatome was officially launched in January 2018 and assigned the name Enfission. Commercial sales of the fuel are expected by 2026.

Lightbridge is working with four US nuclear utilities, and late in 2016 a letter of intent was signed with one of them for a lead test fuel assembly demonstration in a US commercial reactor, possibly by 2021.

Thorium-uranium Fuel under Development

Since the early 1990s Russia has had a program to develop a thorium-uranium fuel, based at Moscow's Kurchatov Institute and involving the US company Lightbridge Corporation and US government funding to design fuel for Russian VVER-1000 reactors. Whereas normal fuel uses enriched uranium oxide throughout the fuel assembly, the new design has a demountable centre portion and blanket arrangement, with uranium-zirconium metal fuel rods in the centre and uranium-thorium oxide pellets in conventional fuel rods around it. The Th-232 in the blanket captures neutrons to become U-233, which is fissile and is fissioned in situ. Blanket material remains in the reactor for nine years but the centre portion is burned for only three cycles (as in a normal VVER, 3 or 4.5 years depending on the refuelling interval). No reprocessing of the blanket is envisaged, due both to the difficulty of doing so with thorium fuels and the presence of significant U-232 in the blanket. The two-part fuel assembly has the same geometry as a normal VVER one.

A variant of this design uses plutonium-zirconium metal fuel rods in the central seed assembly, and was earlier known as a plutonium incinerator.

Other Fuel Developments

Other fuel technologies that seem particularly promising, and which could be commercially deployed in the foreseeable future include:

- Ceramic or coated zirconium claddings that prevent the adverse interaction between steam and zirconium at very high temperature.

- High thermal conductivity oxide fuel, such as can be achieved by including additives like beryllium oxide (BeO). Higher conductivity provides higher safety margins and can allow higher operating powers.
- Thoria-based fuels, including mixed thorium-plutonium (Th-MOX) fuel which can achieve a high utilization factor for recycled plutonium.
- Other all-metal fuels and annular LWR fuel, allowing more cooling and therefore safe, high power densities for the fuel and improved economics.
- Pelletized coated-particle fuels, aimed at achieving high safety levels for the fuel that can be left in a light water reactor for very long periods, thereby achieving high burn-up of recycled plutonium and/or actinide waste components.

Purex Process

Fuel is processed using a hydrometallurgical process known as Purex (Plutonium Uranium Refining by Extraction). This process was originally developed in the USA during the 1940's as a means of recovering plutonium from reactors designed to produce military-grade plutonium with a high fissile isotope 239 content.

Reprocessing is considered to be a proliferating technology, even though plutonium produced in civilian reactors is unsuitable for use in atomic bomb-making. This is because the technique may also be applied to irradiated fuel from specialised reactors that produce military-grade plutonium. Consequently, reprocessing activities are closely monitored, and IAEA and EURATOM cameras constantly track operations.

The aim of this chemical extraction process is to separate irradiated fuel's two main components - uranium and plutonium - from a residue of fission products and minor actinides considered to be waste.

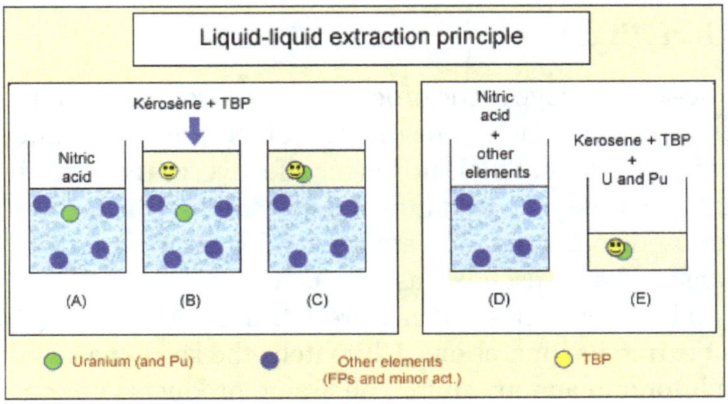

Extraction Phase

The uranium and plutonium (shown in green) are initially mixed with other radioactive elements (shown in dark blue) in strongly acidic water. A kerosene-based oily mixture (light yellow) is added. This mixture contains a special molecule, TriButylPhosphate or TBP (bright yellow), which is only effective in the presence of strongly acidic water. The mixture is agitated (C) and the TBP traps the uranium and plutonium in the kerosene. The acidic water naturally separates from the kerosene when the mixture is allowed to rest. The uranium and plutonium remain trapped in the kerosene. The kerosene containing the uranium and plutonium are then drawn off, leaving behind the acid bath containing the remaining elements.

During shearing operations, irradiated fuel pellets fall into a bath of boiling nitric acid, where they dissolve at a ratio of around 40 kg in approximately 600 litres of acid solution. The resulting solution contains multiple chemical elements, including radioactive and non-radioactive fission products, uranium and plutonium.

A kerosene-based oily mixture containing a special-purpose molecule - TriButylPhosphate or TBP – is then added. TBP, which acts like an enzyme absorbing uranium and plutonium atoms, is only effective in the presence of strongly acidic water. The mixture is agitated - as though making French dressing - and the TBP selectively traps the uranium and plutonium in the kerosene.

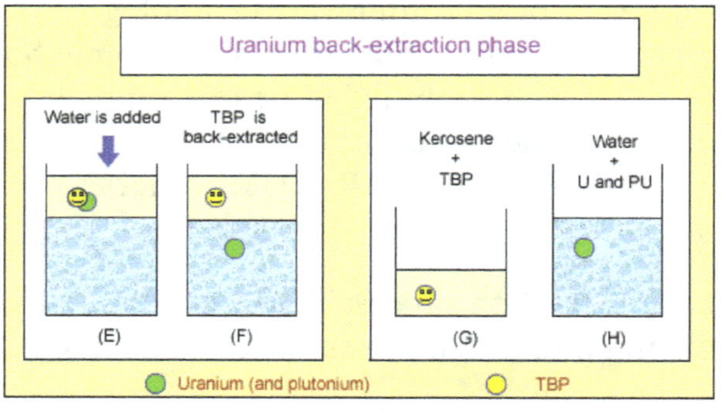

Back-extraction Phase

The purpose of the second stage of the process is to recover the uranium and plutonium from the kerosene. The kerosene is brought into contact with pure water and the resulting mixture agitated as before. TBP is no longer effective, as the water is not sufficiently acidic. As a result, it releases the uranium and plutonium, which enter the pure water. The water and kerosene naturally separate when the mixture is allowed to rest. All that remains is to transfer the kerosene. This results in water containing just the uranium and plutonium, and the kerosene-TBP mixture, which can be reused (almost indefinitely) in subsequent extraction operations. Ultimately, the kerosene solution contains the majority of the plutonium and uranium. The fission products and other transuranium elements remain in the aqueous solution.

The plutonium and uranium are then separated from each other by reducing the plutonium. Several cycles are necessary in order to achieve maximum purification. The uranium, in the form of uranyl nitrate, is purified in several stages by liquid-liquid extraction, removing the fission products. The plutonium is also purified, by extraction, and then concentrated and precipitated as oxalate. The residual solution containing the fission products and actinides is calcined to enable it to be vitrified.

Nuclear Binding Energy

Nuclei are made up of protons and neutrons, but the mass of a nucleus is always less than the sum of the individual masses of the protons and neutrons which constitute it. The difference is a measure of the nuclear binding energy which holds the nucleus together. This binding energy can be calculated from the Einstein relationship:

$$\text{Nuclear binding energy} = \Delta mc^2$$

For the alpha particle $\Delta m = 0.0304$ u which gives a binding energy of 28.3 MeV.

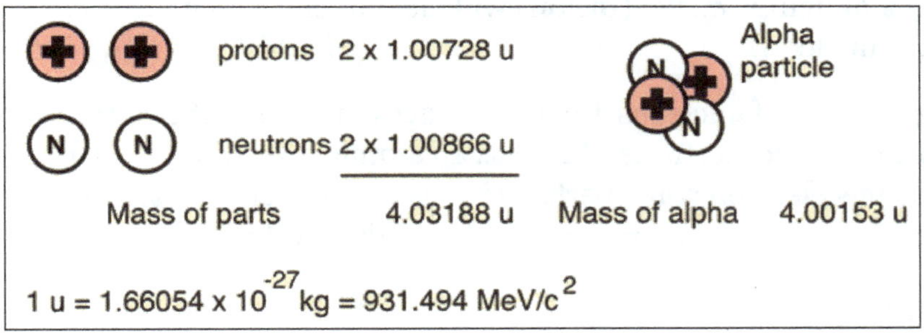

The enormity of the nuclear binding energy can perhaps be better appreciated by comparing it to the binding energy of an electron in an atom. The comparison of the alpha particle binding energy with the binding energy of the electron in a hydrogen atom is shown below. The nuclear binding energies are on the order of a million times greater than the electron binding energies of atoms.

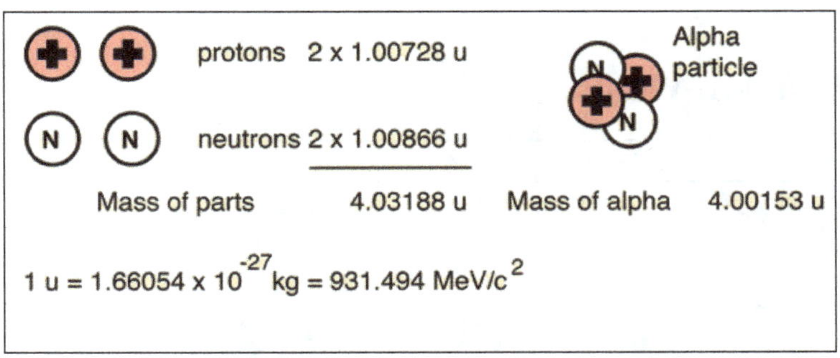

Comparison of atomic and nuclear scales and binding energy.

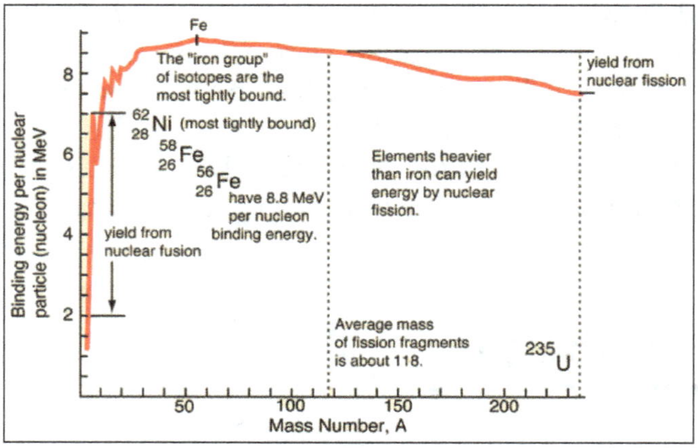

Fission and Fusion can Yield Energy.

Nuclear Binding Energy Curve

The binding energy curve is obtained by dividing the total nuclear binding energy by the number of nucleons. The fact that there is a peak in the binding energy curve in the region of stability near iron means that either the breakup of heavier nuclei (fission) or the combining of lighter nuclei (fusion) will yield nuclei which are more tightly bound (less mass per nucleon).

The binding energies of nucleons are in the range of millions of electron volts compared to tens of eV for atomic electrons. Whereas an atomic transition might emit a photon in the range of a few electron volts, perhaps in the visible light region, nuclear transitions can emit gamma-rays with quantum energies in the MeV range.

The Iron Limit

The buildup of heavier elements in the nuclear fusion processes in stars is limited to elements below iron, since the fusion of iron would subtract energy rather than provide it. Iron-56 is abundant in stellar processes, and with a binding energy per nucleon of 8.8 MeV, it is the third most tightly bound of the nuclides. Its average binding energy per nucleon is exceeded only by ^{58}Fe and ^{62}Ni, the nickel isotope being the most tightly bound of the nuclides.

Fission and Fusion Yields

FUSION	FISSION
deuterium 1875.62 MeV *Fuel* tritium 2808.92 MeV	^{235}U 218,896.8 MeV *Fuel* neutron 939.57 MeV

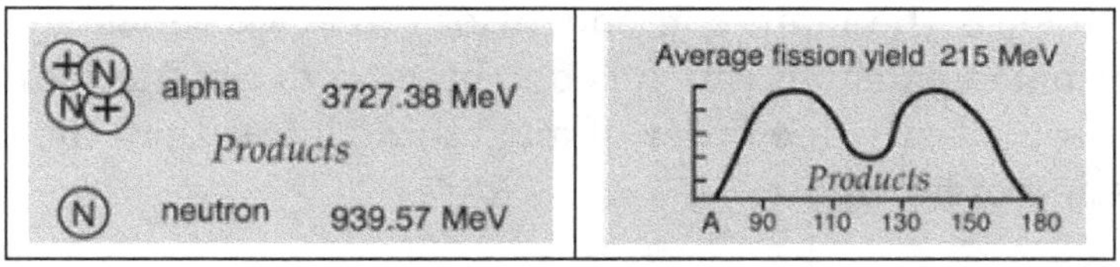

D-T fusion yield: 17.6 MeV

Fractional yield $\dfrac{17.6\,\text{MeV}}{4684.54\,\text{MeV}} = .00375$ $\qquad \dfrac{215\,\text{MeV}}{219{,}836.37\,\text{MeV}} = .00098$

Calculation of the number of reference units of energy produced from 1 kg of fuel by fusion and fission.

3.38×10^{14} per kg of fuel; Δmc^2 8.80×10^{13} J

676   1 UNIT = energy use of one U.S. citizen in 1 year   176

Deuterium-tritium fusion and uranium-235 fission are compared in terms of energy yield. Both the single event energy and the energy per kilogram of fuel are compared. Then they are expressed in terms of a nominal per capita U.S. energy use: 5×10^{11} joules. This figure is dated and probably high, but it gives a basis for comparison. The values above are the total energy yield, not the energy delivered to a consumer.

Fission Products Chemistry

The safety of a nuclear reactor is directly linked to the performance of the nuclear fuel under normal operating and accidental conditions. This nuclear fuel is subjected to extreme conditions of temperatures and gradients, a corrosive environment, and damage from high energy particles from fission. In addition, fission products such as xenon (Xe), krypton (Kr), cesium (Cs), iodine (I), tellurium (Te), molybdenum (Mo), barium (Ba), strontium (Sr) etc, are generated during irradiation, which makes the chemistry of the mixed oxide (MOX) nuclear fuel particularly complex. Some of the fission products are soluble into the fuel matrix, while others form gas, metallic or oxide precipitates depending on conditions of temperature and oxygen potential in the reactor.

Cs, Mo, I and Te in particular are generated with a high fission yield, and subsequently migrate from the centre of the fuel pin towards the pellet edge due to their high volatility, and to the strong temperature gradient (~450 mm^{-1}). They accumulate in the space between fuel and cladding, forming a multi-component phase, the so called "Joint-Oxide Gain" (JOG), in the form of a layer of 150-200 μm. Cesium molybdate Cs_2MoO_4 is the main component of the JOG, which shows a high and anisotropic thermal expansion, which is detrimental to the mechanical properties, and a thermal conductivity one order of magnitude lower than the fuel, which can lead to the creation of hot spots.

The chemistry of cesium compounds forming in the JOG is particularly relevant since the potential release of ^{135}Cs and ^{137}Cs (the isotopes which are the main cause for the radiological consequences of an accident) into the environment is a subject of primary concern for the public. At TU Delft we investigate in the radioactive laboratories of the Reactor Institute Delft the complex chemistry of the JOG combining structure studies by X-ray and neutron diffraction, X-ray Absorption Spectroscopy (XAS), thermodynamic measurements by Differential Scanning Calorimetry (DSC) and solution calorimetry, as well as thermodynamic modelling assessments using the CALPHAD method.

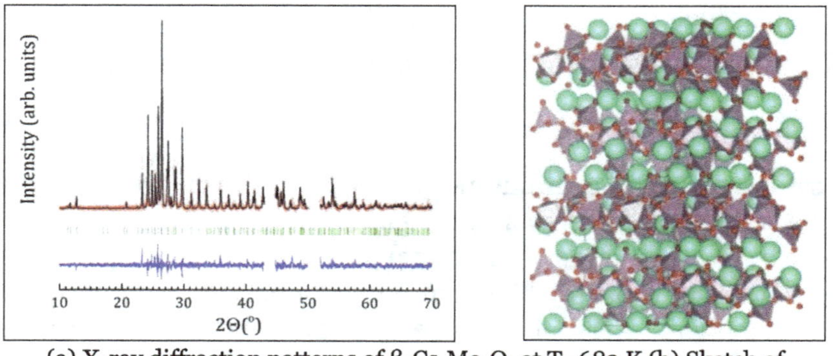

(a) X-ray diffraction patterns of β-$Cs_2Mo_2O_7$ at T=683 K (b) Sketch of the structure of α-$Cs_2Mo_2O_7$ at room temperature.

PWR and BWR Chemistry Optimization

Water chemistry is evolving to cope with the requirements for higher-performance fuel, safe operation with satisfactory long-term component behaviour, options that are economical and compatible with the environment, and finally low radiation exposure.

The challenges of NPP chemistry have changed a lot, although some aspects remain the same. In the past, avoiding the most typical cases of corrosion constituted the main objective. Then pH optimisation of the primary water for dose rate limitation has evolved to a higher pH in the primary coolant; the addition of zinc in primary coolant is now seen as an additional way of decreasing occupational radiation exposure of plant staff. Mitigating crud-induced power shifts (also called axial offset anomaly) along the fuel elements is another challenge with high burn-up fuel and long cycles.

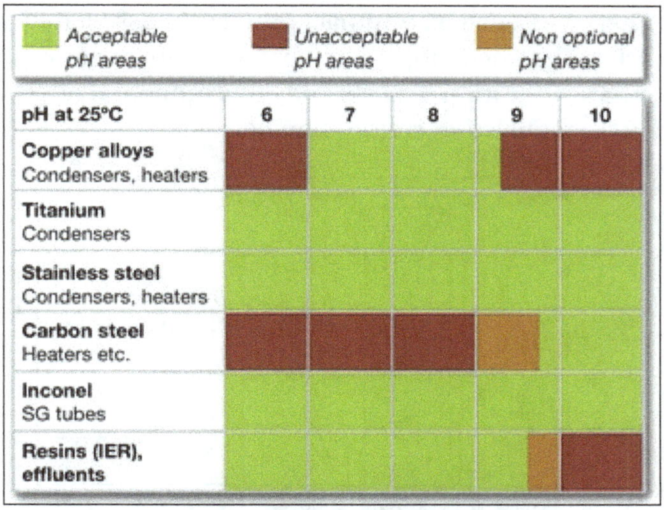

pH restrictions related to the materials.

On the secondary side of the steam generator, the use of Alloy 600, a highly sensitive material, has caused widespread steam generator tubing corrosion that chemistry control was only able to mitigate, but unable to fully avoid. After more than 30 years corrosion-free, the new material, Alloy 690, has demonstrated that the widespread incidence of corrosion was mainly due to a poor material selection (although in some cases, chemistry errors have drastically accelerated the degradation at the origin of early steam generator replacements).

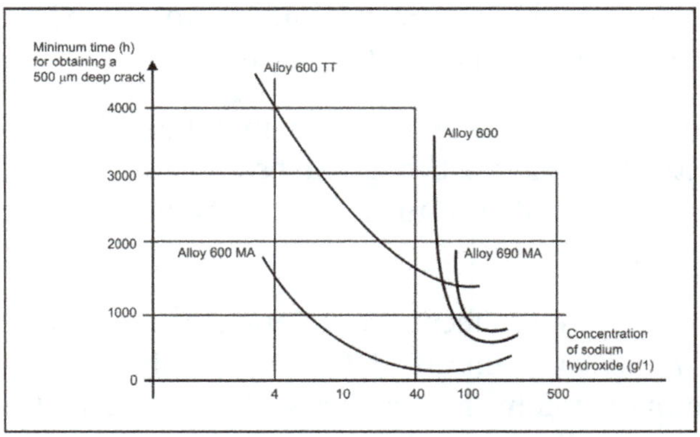

Effect of NaOH concentration on crack initiation by material type.

The new challenges of water chemistry are minimising corrosion of various components of the secondary system, by selecting the best treatment, and maintaining a sufficiently high pH at operating temperature which will minimize corrosion product transport to the steam generator. These oxides may deposit in steam generator tubing (fouling phenomenon) and are at the origin of thermal transfer decrease. They may also increase the risk of flow-induced vibration by modification of the flow within the tube bundle. They build a location where impurities are concentrated so that they may increase the risk of corrosion. Their presence also impacts the costs associated with maintenance needs for chemical cleaning and sludge lancing, which are necessary to eliminate such deposits. Finally, the consideration of costs and environmental effects are taking a larger part in rationale for chemistry selection than before.

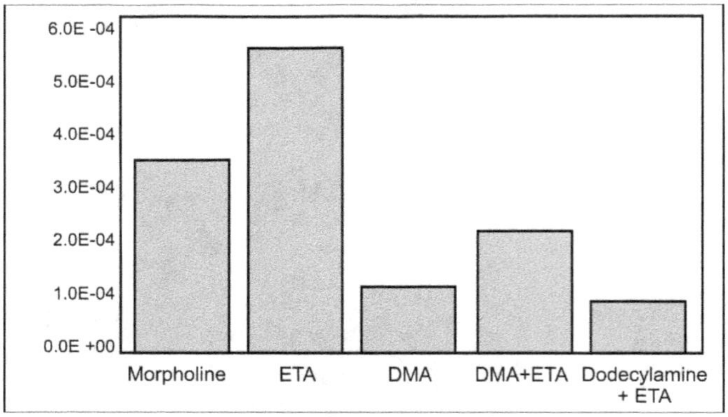

Amine influence on fouling.

PWR Primary Water Chemistry

There are three main tools used to handle reactor coolant system chemistry: pH control, zinc injection and dissolved hydrogen.

Boron-10 concentration is imposed by core characteristics and design and by the fuel burn up. It progressively decreases along the fuel cycle while the fuel burn-up increases. Consequently, the optimum high-temperature pH 300 °C to be maintained requires a decreasing lithium concentration along the fuel cycle.

In the past, a constant pH 300 °C of 6.9 was recommended, because that was the point of minimum solubility of magnetite, thought to be the major component of crud. However, the primary constituent of crud turned out to be nickel ferrite, whose minimum solubility point is at a higher pH.

Then, calculation codes as well as feedback experience agreed to select a pH 300 °C of approximately 7.2 +/- 0.2 for minimizing (i) The solubility and corrosion of the surfaces of the main components in the primary system, (ii) The difference of solubility and thus of transport of radioactivated isotopes in the various parts of the system at different temperatures, typically in the 270 – 325 °C range. However, extending the duration of a fuel cycle to 18 months or more meant that the boron concentration at the beginning of cycle had to be raised from 1200 to 1800 mg/kg (ppm). This requires a lithium concentration greater than 2.2 mg/kg to maintain optimum pH.

However, a limit of 2.2 mg/kg was also imposed to avoid excessive fuel cladding corrosion. Despite these guidelines, field tests and design codes based on solubility and corrosion product transport showed that the optimum pH 300 °C to minimize dose rates and subsequent operational radiation exposure was closer to 7.4. This is why several units started to increase their maximum Li concentration to values ranging from 3 to 6 mg/kg lithium.

But some utilities are still reluctant to increase the Li concentration to values greater than 3 mg/kg for several reasons: (i) The absence of clear benefit of such an increase; (ii) The fuel vendors' restriction for avoiding risk of corrosion of Zircaloy or other Zr alloy; (iii) The risk of stress corrosion cracking (SCC) of the remaining Alloy 600 components in pure water and even more when the alkalinity increases; (iv) The concern for SCC of some stainless steel components in the reactor coolant system. Several utilities concluded that the possible benefits of increasing the pH do not deserve the associated risks of Li concentrations above 3.5 mg/kg, without sufficient proof of the lack of negative consequences.

Another important aspect of water chemistry with growing interest is the addition of zinc. The beneficial influence of Zn has been discovered on shutdown dose rates in boiling water reactor (BWR) units with Cu/Zn alloy condensers. Zinc is now added in the primary water of a growing number of PWR units at a concentration of about 5 µg/kg (ppb), which is sufficient to reduce the dose rates by 15% during the first fuel cycle and up to 30% to 40% after a few cycles with zinc injection.

The beneficial effect is explained by zinc's affinity for the tetrahedral sites of spinel oxide minerals' structure. Thus, it can displace other cations such as iron, or nickel from oxide films, which activates into Co-58, a major radioelement.

In some plants, Zn has been added to mitigate the initiation of primary water stress corrosion cracking on Alloy 600 due to its incorporation in spinel oxide films, enhancing their stability. In this case a higher concentration of Zn is required (25-30 µg/kg), which is not easily accepted by fuel vendors, utilities and regulators due to an insufficient feedback on absence of side effects, on fuel oxide thickness and potential subsequent fuel degradation. Consequently, the application of zinc for primary water SCC (PWSCC) is very limited. Moreover, it's beneficial influence on crack propagation has not been clearly demonstrated, which jeopardizes interest in its addition, particularly if key components such as steam generators have to be replaced soon.

The other controversial chemical parameter is hydrogen concentration, applied to mitigate water radiolysis. For a long time, a value of 25-35 ml/kg (within a range of 25-50) has been used. The US Electric Power Research Institute (EPRI) is considering increasing the hydrogen concentration towards 50 ml/kg and possibly beyond, depending on the results of the ongoing qualification work on the risk of decreasing the time to crack initiation. The intention is to minimise crack growth rates. However, Japanese utilities are evaluating the action of decreasing hydrogen concentration to delay crack initiation.

The preferred strategy for PWRs is to adjust chemistry in a series of small, controlled steps towards the optimum values. The guiding principle is to avoid large changes at once and to monitor progress as the desired option is approached.

Another important concern in PWRs is the mitigation of CIPS (Crud Induced Power Shift) also named AOA (Axial Offset Anomaly) of neutron flux. The International Atomic Energy Agency programme FUWAC will provide relevant data on the phenomenon, taking into account high burnup operation, mixed cores and plant aging.

BWR Primary Water Chemistry

In BWRs, control of stress corrosion cracking has been the prime objective for the past 25 years. Mitigation of the resultant radiation field is a secondary objective. Improvements in water purity reduced the incidence of fuel failures in the late 1980s and 1990s, but margins were reduced as fuel duty increased. In the past, due to fuel failures, tighter limits on water chemistry have been applied to minimize crud build-up. Although these new limits were successful in avoiding any further failures, chemistry specifications continue to be tightened in response to fuel duty increases.

To further optimize plant performance, the nuclear industry has established a number of initiatives to mitigate materials degradation through a proactive approach to existing and potential new corrosion problems, to improve fuel integrity and to enhance radiation protection over the next decade.

Like in PWR plants, BWR coolant chemistry is intended to meet the sometimes conflicting requirements of mitigating intergranular SCC (IGSCC), controlling radiation field and avoiding fuel performance issues. Reduction in feedwater iron ingress and maintaining feedwater zinc has mitigated fuel crud concerns.

Several advanced chemistry improvement steps have been made to assure BWR system integrity: hydrogen injection, zinc addition, NMCA (noble metal chemical addition) and its online process (OLNC). Advanced chemical treatments are applied in all GE-designed BWRs, whereas plants of other manufacturers continued, with good reason, to apply NWC (normal water chemistry). The ANT International recommendation for BWR coolant chemistry is not to use the PWR approach, to not approach the optimum strategy in a series of small steps. Here the prime need is to establish protection of reactor internals by mitigating IGSCC using moderate hydrogen water chemistry or HWC/NMCA. OLNC shows great promise as the most effective way of maintaining long-term mitigation.

PWR/VVER Secondary Water Chemistry

There are several difficulties in the secondary system due to the risk of impurities that may enter through condenser leaks, since this component is cooled by raw water, and its pressure is higher than the secondary system pressure. Impurities will concentrate in the steam generator where there is a large amount of vaporisation. Moreover, restricted flows in some parts of the steam generator can highly concentrate impurities, up to concentration factors of 105. During power operation, impurities will either precipitate on heat transfer walls or will concentrate locally; this is called 'hideout'. During power decrease or shutdown, impurities are partially solubilised and observed at steam generator (SG) blowdown (called 'hideout return').

The main objectives of the secondary water chemistry are:

- Maintain the optimum pH by reagent addition.
- Maintain a reducing environment by limitation of air ingress and hydrazine (N_2H_4) addition.
- Mitigate intergranular corrosion on outer diameter of 600 MA SG tubes by avoiding an alkaline environment and limiting the presence of impurities (alkaline, chloride, sulphate, resin fines, lead, and so on).
- Mitigate flow accelerated corrosion (FAC) of carbon steel and of transport of corrosion products with a sufficiently high pH.
- Reduce the deposition rate inside the SG to maintain thermal heat flux and hydraulic flow; in so doing avoid the risk of impurity concentration and decrease the need for maintenance (sludge lancing, chemical cleaning, and so on).
- Optimise operating costs and application of adequate chemistry specifications and treatment selections to avoid component degradation, unnecessary shutdown and useless operating process with purification systems (for example avoiding a condensate polishing station reduces investment and operating costs).
- Maintain an acceptable level of waste releases into the environment through specification, operational practices, design options, selection of conditioning.

The key points for secondary water chemistry are the risk of impurities and concentration within the SG by the hideout phenomenon; optimisation of the all-volatile treatment (AVT) for minimizing SG fouling and corrosion of the overall system; and minimizing contaminant ingress.

As in the primary coolant, the secondary system (steam-water) pH must be slightly alkaline to reduce component corrosion to a minimum. But in this system, selecting an alkaline reagent is more complicated for various reasons:

- Several materials present, including stainless steel, nickel base alloy and carbon steel are present. Carbon steel is not resistant to FAC if the pH is too low; but also, copper alloys in some plants are very sensitive to corrosion in presence of ammonia or amine if the pH is too high.
- The secondary side experiences a wide temperature range, from about 40 °C in the condenser to 285 °C in the steam generator.

- Steam and liquid phases in the system require the use of volatile alkaline reagents to reach all parts of the circuit.
- Cost and environmental considerations limit the choice of pH and reagent.

Design considerations of the secondary system for selection of optimum pH depending on materials present. All volatile treatment is now the leading method worldwide for secondary water chemistry. There are two main options of what to add: ammonia at pH 25 °C of 10 (high AVT) or an amine (such as morpholine, ethanolamine, DMA, MPA).

Ammonia is easy to use and partially or totally obtained by thermal decomposition of hydrazine. In addition, various amines may be used to get the target pHT. Each reagent has its own advantages and disadvantages for protection of a defined material or system component, and for operational practicality (in terms of issues such as costs, resin life duration and wastes).

It is also necessary to add a sufficient quantity of hydrazine to obtain a reducing environment and minimize risks of SG tubes cracking. However, it is useless and even inappropriate (in terms of costs and health hazards) to add too much hydrazine, as some utilities may have done. If the pH remains too low, with or without amine, it is better to add ammonia rather than increasing hydrazine greater than 100 μg/kg. There is an international consensus for a target value of hydrazine greater than 20 μg/kg and greater than eight times the oxygen concentration. The proposed range by ANT is from 20 to 100 μg/kg in feedwater, with a target value of 50 μg/kg.

Ammonia requires a higher pH 25 °C than morpholine or ETA, since the pH 300 °C will be much lower than with amines because of physical and chemical properties of the reagent, including the evolution of dissociation constants versus temperature and distribution coefficients in steam/water.

Morpholine, widely used in the French fleet of PWRs, has the advantage of a constant distribution coefficient of 1; this translates into a constant concentration throughout the secondary system and thus a good protection of many parts of the system against FAC. However, its limited stability induces the presence of some organic acids, increasing the cation conductivity in the steam generator blowdown and decreasing the detection sensitivity of other impurities.

Compared with other amines, ethanolamine (ETA) has a higher alkalinity, requiring a lower molar concentration. It is also more favourable for the environment, and has a good compatibility with the use of condensate polishers. Consequently it is used in many countries. It also has a better thermal stability than morpholine.

The choice of AVT reagent will depend on the presence or absence of copper alloys, and the presence or absence of condensate polishing plant or a purification system at steam generator blowdown. Ammonia is a viable option for operation at pH 25 °C of 9.9-10 and is easy to implement and monitor, but can only be used in plants without copper alloys. It is normally not used when condensate polishers are in service. Morpholine is a good selection when the condensate polishers are not permanently in operation and when the blowdown resins are not regenerated (to avoid any risk of chemical or resin fines contamination, and to limit liquid chemical effluents). It also provides a uniform protection to all parts of the secondary system against FAC. In addition to French fleet, it has been successfully applied in several VVER units, with an important reduction of corrosion transport and of the number of corroded SG tubes. Ethanolamine (ETA) is a good choice when the condensate polisher is in continuous service or if specific parts of the secondary system must be protected against FAC, or if nitrogen release into the environment must be decreased, or finally if the reagent cost is in favour of ETA.

The main chemistry-related corrosion phenomena that have occurred in the secondary system have been:

- Some pitting of steam generator tubing due to the presence of chloride; stainless steel of Russian design is particularly sensitive.
- Denting of steam generator tubing due to generalized corrosion of carbon steel in the 1970s when switching to AVT.
- Wastage of steam generator tubing due to acidic environment associated with phosphate treatment (mainly German units).
- FAC of carbon steel (mainly feedwater circuit) where the decrease of piping thickness led to some failures, of which the last important incident was at Mihama in 2004, when several people were killed or injured by the pipe burst.
- IGA/SCC (Intergranular Attack, Stress Corrosion Cracking) of Alloy 600 MA in many countries, whatever the applied chemistry.

The remedy for FAC consists of selecting more resistant materials containing chromium, by increasing pH at operating temperature by a proper choice of treatment, or by modifying the flow velocity.

IGA/SCC has been the most important disease of PWR steam generators and has caused many replacements (together with SCC on the primary side, and mechanical degradation). Although IGA/SCC occurs more easily in some chemical environments, its main cause is the existence of a highly sensitive material, Alloy 600. The new material Alloy 690 has not shown any corrosion associated with chemistry after more than 30 years of operation, while Alloy 600 MA has been affected very quickly (even within a few years) in many plants.

Oxidising and alkaline environments are the most conducive to IGA/SCC. The presence of various contaminants, such as sodium as a caustic pollutant, oxygen from air, resin fines, sulphur compounds, and lead, increase the material sensitivity to corrosion. These impurities may easily concentrate when corrosion products (iron oxides) deposit in the steam generator. This deposition phenomenon is why focus is now given to optimizing secondary water treatment to minimize the transport and deposition of such oxides. Impurity control is carried out at steam generator blowdown, where impurities are concentrated and where detection sensitivity is the highest. Typically, the controlled parameters are sodium (which is the cation representative of the most detrimental species, present in most pollutants) and cation conductivity, which is representative of all anionic pollution, that is, Cl, SO_4, organic acids, CO_2.

A last aspect of the chemistry of the secondary system is the evolution of purification systems. When condensers were not leak-tight, many plants had a condensate polishing system (CPS) to eliminate impurities coming from condenser leaks. New materials such as titanium or even stainless steel have now replaced old copper alloys. With these materials, the CPS is more of a burden than an opportunity to eliminate pollution. Laboratory studies and plant experience have shown that decomposition products of resin fines (reduced sulphur compounds), or sodium hydroxide throw (release from the resins), may constitute a very detrimental hazard for SG integrity. Since reliable condensers now exist and low leak detection with helium is carried out, avoiding continuous operation of the condensate polishing system looks to be more desirable.

It appears that feedwater corrosion product transport (iron oxides) is finally higher in units with CPS than those without, as a consequence of pH limitation. CPS operation is hardly compatible with a high pH (regeneration frequency, liquid wastes, and costs) and its benefits do not compensate the inconvenience of these lower pH values. In presence of a CPS, the pH cannot be high enough to minimise the quantity of corrosion products. With a high pH, the frequency of ion exchange resin regeneration becomes unrealistically high in terms of reagent costs, resin degradation, quantity of liquid wastes and associated environmental impact.

Storage and Disposal of Radioactive Waste

- Radioactive wastes are stored so as to avoid any chance of radiation exposure to people, or any pollution.
- The radioactivity of the wastes decays with time, providing a strong incentive to store high-level waste for about 50 years before disposal.
- Disposal of low-level waste is straightforward and can be undertaken safely almost anywhere.
- Storage of used fuel is normally under water for at least five years and then often in dry storage.

- Deep geological disposal is widely agreed to be the best solution for final disposal of the most radioactive waste produced.

Most low-level radioactive waste (LLW) is typically sent to land-based disposal immediately following its packaging for long-term management. This means that for the majority (~90% by volume) of all of the waste types produced by nuclear technologies, a satisfactory disposal means has been developed and is being implemented around the world.

For used fuel designated as high-level radioactive waste (HLW), the first step is storage to allow decay of radioactivity and heat, making handling much safer. Storage of used fuel may be in ponds or dry casks, either at reactor sites or centrally. Beyond storage, many options have been investigated which seek to provide publicly acceptable, safe, and environmentally sound solutions to the final management of radioactive waste. The most widely favoured solution is deep geological disposal. The focus is on how and where to construct such facilities.

Used fuel that is not intended for direct disposal may instead be reprocessed in order to recycle the uranium and plutonium it contains. Some separated liquid HLW arises during reprocessing; this is vitrified in glass and stored pending final disposal.

Intermediate-level radioactive waste (ILW) that contains long-lived radioisotopes is also stored pending disposal in a geological repository. In the USA, defence-related transuranic (TRU) waste – which has similar levels of radioactivity to some ILW – is disposed of in the Waste Isolation Pilot Plant (WIPP) deep geological repository in New Mexico. A number of countries dispose of ILW containing short-lived radioisotopes in near-surface disposal facilities, as used for LLW disposal.

Some countries are at the preliminary stages of their consideration of disposal for ILW and HLW, whilst others, such as Finland and Sweden, have made good progress in the selection of publicly acceptable sites for the future disposal of radioactive waste.

The following table sets out the commonly accepted disposal options. When considering these, it should be noted that the suitability of an option or idea is dependent on the waste form, volume, and radioactivity of the waste.

Table: Commonly-accepted Disposal Options.

Option	Suitable waste types	Examples
Near-surface disposal at ground level, or in caverns below ground level (at depths of tens of metres)	LLW and short-lived ILW	• Implemented for LLW in many countries, including Czech Republic, Finland, France, Japan, Netherlands, Spain, Sweden, UK, and USA. • Implemented in Finland and Sweden for LLW and short-lived ILW.

Deep geological disposal (at depths between 250m and 1000m for mined repositories, or 2000m to 5000m for boreholes)	Long-lived ILW and HLW (including used fuel)	• Most countries have investigated deep geological disposal and it is official policy in several countries. • Implemented in the USA for defence-related truansuranic waste at WIPP. • Preferred sites selected in France, Sweden, Finland, and the USA. • Geological repository site selection process commenced in the UK and Canada.

Near-surface Disposal

The International Atomic Energy Agency (IAEA) definition of this option is the disposal of waste, with or without engineered barriers, in:

- Near-surface disposal facilities at ground level. These facilities are on or below the surface where the protective covering is of the order of a few metres thick. Waste containers are placed in constructed vaults and when full the vaults are backfilled. Eventually they will be covered and capped with an impermeable membrane and topsoil. These facilities may incorporate some form of drainage and possibly a gas venting system.
- Near-surface disposal facilities in caverns below ground level. Unlike near-surface disposal at ground level, where the excavations are conducted from the surface, shallow disposal requires underground excavation of caverns. The facility is at a depth of several tens of metres below the Earth's surface and accessed through a drift.

The term near-surface disposal replaces the terms 'shallow land' and 'ground disposal', but these older terms are still sometimes used when referring to this option.

These facilities will be affected by long-term climate changes (such as glaciation) and this effect must be taken into account when considering safety, as such changes could disrupt these facilities. This type of facility is therefore typically used for LLW and short-lived ILW with half-lives of up to 30 years.

Near-surface disposal facilities at ground level are currently in operation in:

- UK – LLW Repository at Drigg in Cumbria operated by UK Nuclear Waste Management (a consortium led by Washington Group International with Studsvik UK, Serco, and Areva) on behalf of the Nuclear Decommissioning Authority.
- Spain – El Cabril LLW and ILW disposal facility operated by ENRESA.
- France – Centre de l'Aube operated by Andra.

- Japan – LLW Disposal Center at Rokkasho-Mura operated by Japan Nuclear Fuel Limited.
- USA – five LLW disposal facilities: Texas Compact facility near the New Mexico border, operated by Waste Control Specialists; Barnwell, South Carolina operated by Energy Solutions; Clive, Utah (Energy Solutions); Oak Ridge, Tennessee (Energy Solutions); and Richland, Washington – operated by American Ecology Corporation.

Near-surface disposal facilities in caverns below ground level are currently in operation in:

- Sweden – the SFR final repository for short-lived radioactive waste at Forsmark, where the depth of the facility is 50m under the Baltic seabed – operated by the Swedish Nuclear Fuel and Waste Management Company (SKB).
- Finland – an underground repository at Olkiluoto for LLW and ILW has been in operation since 1992. A similar facility at Loviisa was commissioned in 1997. The depth of these is about 100 metres.

Deep Geological Disposal

The long timescales over which some waste remains radioactive has led to the idea of deep disposal in underground repositories in stable geological formations. Isolation is provided by a combination of engineered and natural barriers (rock, salt, clay) and no obligation to actively maintain the facility is passed on to future generations. This is often termed a 'multi-barrier' concept, with the waste packaging, the engineered repository, and the geology all providing barriers to prevent the radionuclides from reaching humans and the environment. In addition, deep groundwater is generally devoid of oxygen, minimising the possibility of chemical mobilisation of waste.

Deep geological disposal is the preferred option for nuclear waste management in several countries, including Argentina, Australia, Belgium, Canada, Czech Republic, Finland, France, Japan, the Netherlands, Republic of Korea, Russia, Spain, Sweden, Switzerland, the UK, and the USA. Hence, there is much information available on different disposal concepts; a few examples are given here. The only purpose-built deep geological repository that is currently licensed for disposal operations is the Waste Isolation Pilot Plant (WIPP) in the USA. Plans for disposal of spent fuel are well advanced in Finland, Sweden, France, and the USA, though in the USA there have been political delays. In Canada and the UK, deep disposal has been selected and the site selection process has commenced.

Mined Repositories

The most widely proposed deep geological disposal concept is for a mined repository comprising tunnels or caverns into which packaged waste would be placed. In some cases (*e.g.* wet rock) the waste containers are then surrounded by a material such as cement or clay (usually bentonite) to provide another barrier (called buffer and/or backfill). The choice of waste container materials and design, as well as the buffer/ backfill material varies depending on the type of waste to be contained and the nature of the host rock-type available.

Excavation of a deep underground repository using standard mining or civil engineering technology is limited to accessible locations (*e.g.* under land or nearshore), to rock units that are reasonably stable and without major groundwater flow, and to depths of between 250m and 1000m. The contents of the repository would be retrievable in the short term, and if desired, longer-term.

The Swedish proposed KBS-3 disposal concept uses a copper container with a steel insert to contain the spent fuel. After placement in the repository about 500 metres deep in the bedrock, the container would be surrounded by a bentonite clay buffer to provide a very high level of containment of the radioactivity in the spent fuel over a very long time period. In June 2009, the Swedish Nuclear Fuel and Waste Management Company (SKB) announced its decision to locate the repository at Östhammar.

Finland's repository programme is also based on the KBS-3 concept. Spent nuclear fuel packed in copper canisters will be embedded in the Olkiluoto bedrock at a depth of around 400 metres. The country's nuclear waste management company, Posiva Oy, expects the repository to begin disposal operations in 2023. Its construction was licensed in November 2015.

The deposits of native (pure) copper in the world have proven that the copper used in the final disposal container can remain unchanged inside the bedrock for extremely long periods, if the geochemical conditions are appropriate (low levels of groundwater flow). The findings of ancient copper tools, many thousands of years old, also demonstrate the long-term corrosion resistance of copper, making it a credible container material for long-term radioactive waste storage.

Deep Boreholes

As well as mined repositories, which have been the focus of most international efforts so far, deep borehole disposal has been considered as an option for geological isolation for many years, including original evaluations by the US National Academy of Sciences in 1957 and more recent conceptual evaluations. In contrast to recent thinking on mined repositories, the contents would not be retrievable.

The concept consists of drilling a borehole into basement rock to a depth of up to about 5000 metres, emplacing waste canisters containing used nuclear fuel or vitrified radioactive waste from reprocessing in the lower 2000 metres of the borehole, and sealing the upper 3000 metres of the borehole with materials such as bentonite, asphalt or concrete. The disposal zone of a single borehole could thus contain 400 steel canisters each 5 metres long and one-third to half a metre in diameter. The waste containers would be separated from each other by a layer of bentonite or cement.

Boreholes can be readily drilled offshore as well as onshore in both crystalline and sedimentary host rocks. This capability significantly expands the range of locations that can be considered for the disposal of radioactive waste.

Deep borehole concepts have been developed (but not implemented) in several countries, including Denmark, Sweden, Switzerland, and the USA. Compared with deep geological disposal in a mined underground repository, placement in deep boreholes is considered to be more expensive for large volumes of waste. This option was abandoned in countries such as Sweden, Finland, and the USA, largely on economic grounds. The borehole concept remains an attractive proposition for the disposal of smaller waste forms including sealed radioactive sources from medical and industrial applications.

An October 2014 US Department of Energy (DOE) report said: "Preliminary evaluations of deep borehole disposal indicate a high potential for robust isolation of the waste, and the concept could offer a pathway for earlier disposal of some wastes than might be possible in a mined repository." In January 2016 the DOE commissioned a team led by Battelle to drill a 4880-metre test borehole into crystalline basement rock in North Dakota.

Interim Waste Storage and Transport

Specially designed interim surface or sub-surface storage waste facilities are currently used in many countries to ensure the safe storage of hazardous radioactive waste pending the availability of a long-term disposal option. Interim storage facilities are generally used for ILW and HLW, including used nuclear fuel from reactors.

Storage Ponds

Storage ponds at reactors, and those at centralised facilities such as CLAB in Sweden, are 7-12 metres deep to allow the racked fuel assemblies to be covered by several metres of water. The fuel assemblies are typically about 4 m long and standing on end. The multiple racks are made of metal with neutron absorbers incorporated in it. The circulating water both shields and cools the fuel. These pools are robust constructions made of thick reinforced concrete with steel liners. Ponds at reactors may be designed to hold all the used fuel for the life of the reactor, but usually the design assumes some removal of cooled fuel for reprocessing or to dry storage.

Dry Storage

Some storage of fuel assemblies which have been cooling in ponds for at least five years is in dry casks or vaults, typically with air circulation inside concrete shielding. Dry storage has been used at US nuclear power plants since 1986, and about one-third of the total US used fuel is now in dry storage casks. Facilities are at most of the 65 nuclear power plant sites (including some closed ones). Transfer from wet storage to dry casks at a power plant site may use special shielded transfer casks, which are less robust than those used for transport beyond the site. Casks may contain a sealed canister which can be transferred from one kind of cask to another.

Multi-purpose Canisters

Sealed multi-purpose canisters (MPCs), also called dual-purpose canisters (DPCs), each holding up to 89 fuel assemblies with inert gas, are commonly used for transporting, storing and eventual disposal of used fuel. MPCs are contained inside robust overpacks – metal for transport, or mainly concrete for storage. Each MPC, constructed using 13 mm welded stainless steel with a secure lid and internal fuel basket to hold and keep the fuel assemblies separate, is designed for up to 45 kW heat load. MPCs have standard external dimensions and the number of fuel assemblies actually loaded into one depends on their characteristics. Some are double-walled (DWC), with helium in between the layers. Once an MPC is loaded the contents should never need to be handled again.

Holtec's MPC contains a 68-cell fuel basket for BWR fuel, a 24-cell flux-trap, or a 32-cell non-flux trap fuel basket for PWR fuel. Some have neutron-absorbing Metamic-HT fuel baskets and liners with high thermal conductivity, enabling relatively-hot three-year-old spent fuel to be placed in them. Since 2013 Holtec also makes double-walled canisters (DWCs) for the UK and Ukraine, with the same standard external dimensions.

High-capacity MPCs are able to hold 37 PWR or 87-89 BWR fuel assemblies. Others take only 12 PWR or 32 BWR fuel assemblies. With some small canisters – 4 PWR/9BWR – four can be fitted into a standard transport cask. Areva has a large canister design holding 21 PWR or 44 BWR fuel assemblies.

Storage Casks and Systems

For storage, each MPC is enclosed in a ventilated storage module or overpack made of concrete and steel. These are commonly standing on the surface, about 6 metres high, cooled by air convection, or they may be horizontal in banks, or vertical and below grade, with just the tops showing. The modules are robust and provide full shielding. At a nuclear power plant, a shielded transfer cask is used to move the MPC from the used fuel pool to a storage module. Holtec's transfer casks for onsite use are called HI-TRAC.

A free-standing above-ground system is Holtec's HI-STORM 100, which accommodates a variety of sealed stainless steel MPCs vertically inside ventilated concrete and steel overpacks standing on a concrete pad. The 165-tonne overpack has 65 cm of concrete inside steel casing for shielding. This system is used at many US plants. A below-ground variant is HI-STORM 100U, and a more sophisticated version of this is Holtec's HI-STORM

UMAX storage system, which is already deployed at two US nuclear power plant sites, and proposed for the HI-STORE CIS facility in New Mexico. HI-STORM UMAX stores the canisters containing used fuel in ventilated vertical steel and concrete cavity enclosure containers 5 metres high below ground, with massive lids at ground level. The containers are set up in a 7.6 m deep excavation and low-strength concrete grout is backfilled around them. The final half metre of fill is a reinforced concrete pad. Seismic tolerance is about 2000 Gal.

An above-ground horizontal system is Areva's NUHOMS HSM-H system, used by 20 of the 67 ISFSIs at US nuclear plants. The used fuel is sealed in 15 mm thick stainless steel dry storage canisters which are then moved in horizontal shielded transfer casks to large horizontal storage modules, with each hole 5 m long and 2 m diameter. Areva claims better heat distribution in these than vertical systems, using conduction more than convection, and also points to no gaps between modules being very safe seismically (1500 Gal) and radiologically. Each canister holds 32-37 PWR fuel assemblies or 61-69 BWR assemblies, in helium. Areva plans to have a US-licensed transport cask for NUHOMS by 2020, and meanwhile it has one for high burn-up fuel. NUHOMS horizontal storage can accommodate a variety of canister designs, and the Areva NUHOMS canisters are compatible with other, *e.g.* Holtec, vertical storage systems.

In July 2017 the US Nuclear Regulatory Commission granted Areva TN certification for its NUHOMS Extended Optimized Storage (EOS) dry used fuel storage system. The system is designed for high burn-up fuel management and reactors in shutdown phase. The large NUHOMS EOS dry shield canisters hold 37 PWR or 89 BWR fuel assemblies. These can be transferred in the EOS TC series transfer cask and stored in NUHOMS EOS HSM concrete modules.

Germany's Gesellschaft für Nuklear-Service mbH (GNS), set up in 1977 and owned by the country's four nuclear utilities, is both an operator of waste storage and supplier of two types of cask. The main type are CASTOR casks, more than 1000 of which are used in Germany alone for transport and interim storage of spent fuel and HLW. The cask body provides full shielding and allows loading after very short cooling times of high burn-up fuel, and they are closed with two lids. GNS's CONSTOR casks are similar but with concrete in the walls and designed for cooler fuels.

A large canister HI-STORM FW storage module is a flood- and wind-resistant version with high seismic tolerance – 1.2 Gal. It holds 37 PWR assemblies, or 89 BWR ones, or 31 VVER-1000 ones, with maximum heat load 46 kW. The HI-STORM 190 is the VVER version of the HI-STORM FW. These canisters can hold damaged fuel in special failed fuel containers.

A new large storage cask is HI-STORM MIC (mega-impact capable) designed with EDF Energy in the UK and having a 100-year design life. It uses a double-walled MPC, and is heavily shielded.

A collection of casks or modules comprises an independent spent fuel storage installation (ISFSI), which in the USA is licensed separately from any associated power plant, and is for interim storage only. About one-third of US used fuel is stored thus.

Recognising that long-term management options, specifically for ILW and HLW, may require significant time to be achieved, interim storage arrangements may need to be extended beyond the time periods originally envisaged.

An indication of the scale of dry storage is that the market for dry storage casks in the Americas, where there is the highest demand, was over $600 million in 2015 and is expected to rise to over $1 billion by 2020.

Some countries, including Australia, Belgium, the Netherlands, Germany, Italy, and Switzerland also place LLW in interim storage, although most LLW is typically sent directly to land-based near-surface disposal facilities.

ISFSI Systems in use

Many US nuclear power plants, both operational and decommissioned, have independent spent fuel storage installations (ISFSIs).

In the USA a large ISFSI is proposed in Texas, using Areva's NUHOMS system for up to 40,000 tonnes of used fuel. Waste Control Specialists applied for a licence for the facility in April 2016, but progress with this depends on corporate ownership change. Another proposed facility is the HI-STORE consolidated interim storage (CIS) site in New Mexico for the Eddy Lea Energy Alliance, with an initial capacity of 10,000 storage canisters for about 120,000 tonnes of used fuel. This will occupy about 15 hectares. A licence application for HI-STORE CIS was submitted in March 2017.

Zwilag's ZZL in Switzerland and Ahaus and Gorleben in Germany are examples of operating long-term above-ground central interim dry storage for HLW.

Ukraine's new Central Spent Fuel Storage Facility (CSFSF) will use a Hi-STORM 190 ventilated storage system and Hi-STAR 190 casks for transport to the site. Ukraine requires storage to be in double-wall canisters (DWCs), and these are also used at Sizewell in the UK, giving an increased service life. In Ukraine, Holtec is establishing a centre for supply of dry storage and transport systems to the country's 15 operating reactors and possibly other VVERs around the world.

Lithuania is using GNS CONSTOR M2 dry storage containers at Ignalina, each holding about 90 RBMK fuel assemblies. Bulgaria is also using 270 CONSTOR casks for VVER-440 fuel. CONSTOR casks are robust steel casks with concrete in the walls and three lids, cost-optimised for interim storage of spent fuel which has cooled. They are portable though evidently not designed for shipping used fuel. They are not ventilated and may have a concrete overpack.

Japan's Federation of Electric Power Companies of Japan (FEPC) aims to develop an all-aluminum, thinner and lighter basket for a concrete cask that would be capable of storing more spent fuel assemblies at temperatures of up to 400 °C. It wants to have the design ready for deployment by the mid-2020s. Most dry storage casks require fuel to be cooled in ponds until the temperature drops to about 250 °C.

Transport Casks

Used fuel assemblies are shipped in 'Type B' casks which are shielded with steel, or a combination of steel and lead, and can weigh up to 110 tonnes when empty. They contain their highly radioactive payload safely during transport, and may hold from 6 to 24 tonnes of used fuel. Some information on transport of used fuel is given above, because sometimes very similar equipment is used for storage.

For transport beyond the power plant site, the MPC is put inside a steel transport cask such as a version of the Holtec HI-STAR (STAR = storage, transport and repository). This is classified as a Type B shipping cask, providing secure shielding and if necessary also some heat dissipation. IAEA SSR-6 regulations specify a maximum of 10 mSv/hr dose rate at the external surface, though many customers such as CNNC specify 2 mSv/h (these represent measured dose rates of 0.1 and 0.02 mSv/h respectively one metre from the surface).

The internal structure of transport casks (using multi-purpose canisters or not) is designed to maintain separation of fuel assemblies even in extreme accidents, and the external structure is designed to maintain safe containment in extreme accidents. Both features are tested before licensing.

In the USA the Nuclear Regulatory Commission noted about 2013: "Over the last 40 years, thousands of shipments of commercially generated spent nuclear fuel have been made throughout the United States without causing any radiological releases to the environment or harm to the public." Most of these shipments are between different power plants owned by the same utility, so as to share storage space for used fuel.

Holtec produces a number of Type B shipping packages or casks. Its HI-STAR 100 is a high-capacity system which is engineered to accept one sealed multi-purpose canister (MPC) containing a 68-cell fuel basket for BWR fuel or a 32-cell fuel basket for PWR fuel. It has been licensed since 1998 for transport (horizontal) and storage (vertical, on end). The MPC containing the fuel can be transferred to HI-STORM 100 storage systems on an ISFSI pad or below ground surface, exchanging one overpack for another which is engineered for maximum shielding. The MPC is loaded under water. In 2018 Holtec sold its HI-STAR 100MB casks to CNNC Everclean in China – a medium-weight version of the cask for bare fuel or fuel in canisters.

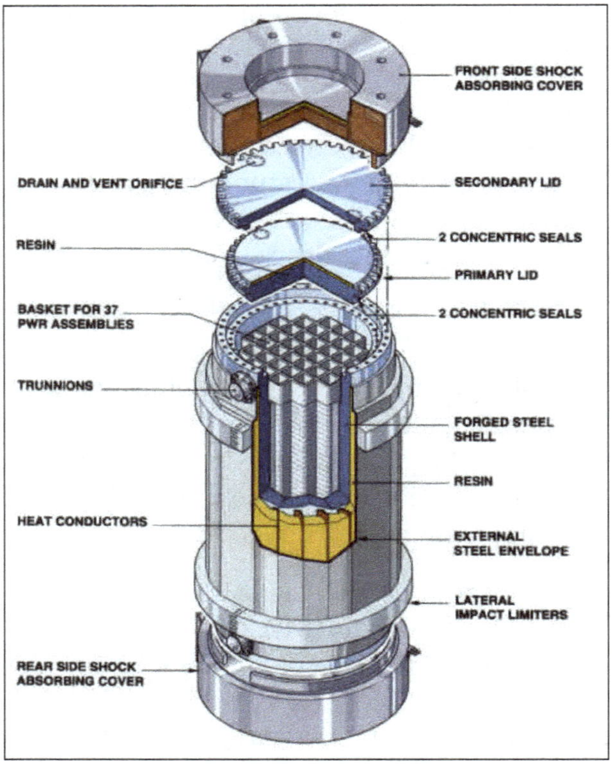

TN24 cask produced by Orano TN (formerly Areva TN).

Areva has a range of Type B transport casks for used fuel, its newest ones being TN112 and TN117. Its older TN12/2 cask designed for used fuel has been adapted for fresh MOX fuel, and holds 12 PWR fuel assemblies or 32 BWR ones. It is robust, with shock-absorbing covers at each end. Each year Areva moves about 200 casks with used fuel and about 20 with vitrified HLW. Fresh MOX fuel and vitrified HLW use different casks than for used fuel. Used MOX fuel requires greater neutron shielding, such as provided by the TN112 cask, which has a double containment barrier and can also take 50 kW heat load. It has lead shielding with steel, while the older one is simply forged steel. Both have resin compound shielding. They are both loaded under water.

Holtec's HI-STAR 80 cask, a multi-layered steel cylinder which holds 12 PWR or 32 BWR high-burnup used fuel assemblies (above 45 GWd/t) which have had cooling times as short as 18 months. The HI-STAR 60 cask can transport 12 PWR used fuel assemblies. The HI-STAR 180 was the first one licensed to transport high burn-up fuel, and holds 32 or 37 PWR used fuel assemblies. The HI-STAR 190 cask has 38 kW heat load capacity and one version is to be used domestically in Ukraine for PWR fuel. It is envisaged as the main used fuel transport to central storage or disposal sites in the USA and is promoted as a universal transport cask. It will accommodate all types of spent fuel MPCs used in the USA (namely Areva, Pacific

Nuclear, Vectra, NAC, Sierra Nuclear, BNFL Solutions, and Westinghouse). All HI-STAR casks have an aluminium impact limiter at each end, making the whole unit about 7 metres long.

The initial licence by the US Nuclear Regulatory Commission covers transport of Holtec's MPC-37 canisters for PWR fuel and MPC-89 canisters for BWR fuel, while Ukraine's SNRIU has certified VVER canisters MPC-31 and MPC-85 for transport in the cask.

In Russia, TUK (transportation packaging set) casks are used to transport used nuclear fuel. Several TUK-13 casks fit into a container or TK carrier for rail transport, each cask holding 12 VVER fuel assemblies (about 6 tonnes). A larger TUK-1410 cask from the Federal Centre for Nuclear & Radiation Safety has now been licensed to replace the older model for VVER-1000 fuel, both in Russia and from overseas. Each weighs over 100 tonnes, holds 18 VVER fuel assemblies weighing 9 tonnes in a removable canister, and is designed for hotter fuel – up to 36 kW heat load. The TUK-14 series is also more heavily shielded, and can carry VVER-1200 fuel. It fits a TK-U-141 railway carrier.

OMZ subsidiary Izhorskiye Zavody has completed the prototype of a new generation of container for the transport and storage of used nuclear fuel from VVER-1000 and VVER-1200 reactors. The TUK-151 container is a thick-walled 2.5m diameter vessel with a sealed lid, a length of about 6 metres and a laden weight of 116 tonnes. Unlike the TUK-13 container design, the TUK-151 is not made of stainless steel, but has a low alloy steel-plated interior surface that offers "superior strength" and resistance to fracturing at low temperatures under dynamic loads.

In the UK, 47- or 53-tonne rectangular Type B flasks have long been used to transport Magnox and AGR fuel, which is held in internal skips. In Sweden, more than 80 large transport casks are shipped annually to a central interim waste storage facility called CLAB Each 80 tonne cask has steel walls 30 cm thick and holds 17 BWR or 7 PWR fuel assemblies. The used fuel is shipped to CLAB after it has been stored for about a year at the reactor, during which time heat and radioactivity diminish considerably. Both these systems are for transport only, not storage.

Spain's ENSA supplied an ENUN 24P dual-purpose cask to CNNC Everclean in China in 2018, for transporting used fuel from Daya Bay inland to Lanzhou in Gansu province. The ENUN cask is licensed for high burn-up fuel, to 57 GWd/tU and, with an external diameter of only 3.3 m, it holds 24 PWR fuel assemblies. The ENUN 32P cask has been used in Spain.

Other Ideas for Disposal

Numerous options for long-term nuclear waste management have been considered in the past. The table below highlights a number of these.

Ideas	Examples
Long-term above ground storage	• Investigated in France, Netherlands, Switzerland, UK, and USA. • Not currently planned to be implemented anywhere.
Disposal in outer space (proposed for wastes that are highly concentrated)	• Investigated by USA. • Investigations now abandoned due to cost and potential risks of launch failure.
Rock-melting(proposed for wastes that are heat-generating)	• Investigated by Russia, UK, and USA. • Not implemented anywhere. • Laboratory studies performed in the UK.
Disposal at subduction zones	• Investigated by USA. • Not implemented anywhere. • Not permitted by international agreements.
Sea disposal	• Implemented by Belgium, France, Germany, Italy, Japan, Netherlands, Russia, South Korea, Switzerland, UK, and USA. • Not permitted by international agreements.
Sub seabed disposal	• Investigated by Sweden and UK (and organisations such as the OECD Nuclear Energy Agency). • Not implemented anywhere. • Not permitted by international agreements.
Disposal in ice sheets (proposed for wastes that are heat-generating)	• Investigated by USA. • Rejected by countries that have signed the Antarctic Treaty or committed to providing solutions within national boundaries.
Deep well injection (for liquid wastes)	• Implemented in Russia for many years for LLW and ILW. • Investigations abandoned in the USA in favour of deep geological disposal of wastes in solid form.

Long-term above Ground Storage

Above ground storage is normally considered an interim measure for the management of radioactive waste. But it has, in the past, been considered as a disposal option. France investigated it for HLW within the framework of the 1991 law on research into radioactive waste management (Act No 91-1381 of 30 December 1991, also known as the 'Bataille Act' after the name of its proposer), but not as a means of final disposal. However, controlled surface storage over longer time periods has also been suggested as a long-term waste management option.

Long-term above ground storage involves specially constructed facilities at the Earth's surface that would be neither backfilled nor permanently sealed. Hence, this option would allow monitoring and retrieval at any time without excessive expenditure.

Suggestions for long-term above ground storage broadly fall into two categories:

- Conventional stores of the type currently used for interim storage, which would require replacement and repackaging of waste every 200 years or so.

- Permanent stores that would be expected to remain intact for tens of thousands of years. These structures are often referred to as 'Monolith' stores or 'Mausoleums'.

The latter category of store is derived from the principle of 'guardianship', where future generations continue to monitor and supervise the waste.

Both suggestions would require information to be passed onto future generations, leading to the question of whether the stability of future societies could be ensured to the extent necessary to continue the required monitoring and supervision.

No country is currently planning to implement long-term (*i.e.* greater than a few hundred years) above ground storage. However, France is investigating long-term interim storage, but not necessarily above ground.

Long-term above ground storage has been considered as part of the range of management concepts in Switzerland by EKRA (Expert Group on Disposal Concepts for Radioactive Waste). EKRA observed that it was unclear what additional steps would be necessary to show how the long-term above ground storage concept could be brought to the state of development comparable with that of geological disposal, and it recommended geological disposal as the preferred option.

Disposal in Outer Space

The objective of this option is to remove the radioactive waste from the Earth, for all time, by ejecting it into outer space. The waste would be packaged so that it would be likely to remain intact under most conceivable accident scenarios. A rocket or space shuttle would be used to launch the packaged waste into space. There are several ultimate destinations for the waste which have been considered, including directing it into the Sun.

The high cost means that such a method of waste disposal could only be appropriate for separated HLW – *i.e.* long-lived highly radioactive material that is relatively small in volume – rather than spent fuel. The question was investigated in the USA by NASA in the late 1970s and early 1980s. Because of the high cost of this option and the safety aspects associated with the risk of launch failure, it was abandoned.

Rock Melting

The deep rock melting option involves the melting of wastes in the adjacent rock. The idea is to produce a stable, solid mass that incorporates the waste, or encases the waste in a diluted form (*i.e.* dispersed throughout a large volume of rock), and that cannot easily be leached and transported back to the surface. This technique has been mainly suggested for heat-generating wastes such as vitrified HLW and host rocks with suitable characteristics to reduce heat loss.

The HLW in liquid or solid form could be placed in an excavated cavity or a deep borehole. The heat generated by the wastes would then accumulate resulting in temperatures great enough to melt the surrounding rock and dissolve the radionuclides in a growing sphere of molten material. As the rock cools it would crystallise and incorporate the radionuclides in the rock matrix, thus dispersing the waste throughout a larger volume of rock. There are some variations of this option in which the heat-generating waste would be placed in containers and the rock around the container melted. Alternatively, if insufficient heat is generated the waste would be immobilised in the rock matrix by conventional or nuclear explosion.

Rock melting has not been implemented anywhere for radioactive waste. There have been no practical demonstrations of the feasibility of this option, apart from laboratory studies of rock melting. In the late 1970s and early 1980s, the rock melting option at depth was taken forward to the engineering design stage. This design involved a shaft or borehole which led to an excavated cavity at a depth of 2.5 kilometres. It was estimated, but not demonstrated, that the waste would be immobilised in a volume of rock 1000 times larger than the original volume of waste.

Another early proposal was for the heat-generating wastes to be emplaced in weighted, heat-resistant containers such that they would melt the underlying rock, allowing them to move downwards to greater depths with the molten rock solidifying above. This proposed option resembles similar self-burial methods proposed for disposal of HLW in ice sheets.

In the 1990s there was renewed interest in this option, particularly for the disposal of limited volumes of specialised HLW (particularly plutonium) in Russia and in the UK. A scheme was proposed in which the waste content of the container, the container composition, and the placement layout would be designed to preserve the container and prevent the wastes becoming incorporated in the molten rock. The host rock would be only partially melted and the container would not move to greater depths.

Russian scientists have proposed that HLW, particularly excess plutonium, could be placed in a deep shaft and immobilised by nuclear explosion. However, the major disturbance to the rock mass and groundwater by the use of nuclear explosions, as well as arms control considerations, has led to the general rejection of this option.

Disposal at Subduction Zones

Subduction zones are areas where one denser section of the Earth's crust is descending beneath another lighter, more buoyant section. The movement of one section of the Earth's crust below another is marked offshore by a trench, and earthquakes commonly occur adjacent to the inclined contact between the two plates. The edge of the overriding plate is crumpled and uplifted to form a mountain chain parallel to the trench. Deep sea sediments may be scraped off the descending slab and incorporated into the adjacent mountains. As the oceanic plate descends into the hot mantle, parts of it may begin to melt. The magma thus formed migrates upwards, some of it reaching the surface as lava erupting from volcanic vents. The idea for this option would be to dispose of wastes in the trench region such that they would be drawn deep into the Earth.

Although subduction zones are present at a number of locations across the Earth's surface, they are geographically very restricted. Not every waste-producing country would be able to consider disposal to deep-sea trenches, unless international solutions were sought. However, this option has not been implemented anywhere and, as it is a form of sea disposal, it is therefore not permitted by international agreements.

Sea Disposal

Disposal at sea involves radioactive waste being dropped into the sea in packaging designed to either: implode at depth, resulting in direct release and dispersion of radioactive material into the sea; or sink to the seabed intact. Over time the physical containment of containers would fail, and remaining radionuclides would be dispersed and diluted in the sea. Further dilution would occur as the radionuclides migrated from the disposal site, carried by currents. The amount of radionuclides remaining in the seawater would be further reduced both by natural radioactive decay, and by the removal of radionuclides to seabed sediments by the process of sorption. This method is not permitted by a number of international agreements.

The application of the sea disposal of LLW and ILW has evolved over time from being a disposal method that was actually implemented by a number of countries, to one that is now banned by international agreements. Countries that have at one time or another undertaken sea disposal using the above techniques include Belgium, France, Germany, Italy, the Netherlands, Sweden, Switzerland, and the UK, as well as Japan, South Korea, and the USA. This option has not been implemented for HLW.

Sub-seabed Disposal

For the sub-seabed disposal option, radioactive waste containers would be buried in a suitable geological setting beneath the deep ocean floor. This option has been suggested for LLW, ILW, and HLW. Variations of this option include:

- A repository located beneath the seabed. The repository would be accessed from land, a small uninhabited island, or from an offshore structure.
- Burial of radioactive waste in deep ocean sediments.

Sub-seabed disposal has not been implemented anywhere and is not permitted by international agreements. The disposal of radioactive wastes in a repository constructed below the seabed has been considered by Sweden and the UK. In comparison to disposal in deep ocean sediments, if it were desirable the repository design concept could be developed so as to ensure that future retrieval of the waste remained possible. The monitoring of wastes in such a repository would also be less problematic than for other forms of sea disposal.

Burial of radioactive waste in deep ocean sediments could be achieved by two different techniques: penetrators or drilling placement. The burial depth of waste containers below the seabed can vary between the two methods. In the case of penetrators, waste containers could be placed about 50 metres into the sediments. Penetrators weighing a few tonnes would fall through the water, gaining enough momentum to embed themselves into the sediments. A key aspect of the disposal of waste to seabed sediments is that the waste is isolated from the seabed by a thickness of sediments. In 1986, some confidence in this process was obtained from experiments undertaken at a water depth of approximately 250 metres in the Mediterranean Sea. The experiments provided evidence that the entry paths created by penetrators were closed and filled with remoulded sediments of about the same density as the surrounding undisturbed sediments.

Wastes could also be placed using drilling equipment based on the techniques in use in the deep sea for about 30 years. By this method, stacks of packaged waste would be placed in holes drilled to a depth of 800 metres below the seabed, with the uppermost container about 300 metres below the seabed.

In the 1980s, the feasibility of the disposal of HLW in deep ocean sediments was investigated and reported by the Organisation for Economic Co-operation and Development (OECD). For this concept, radioactive waste would be packaged in corrosion-resistant containers or glass, which would be placed beneath at least 4000 metres of water in a stable, deep seabed geology chosen both for its slow water flow and for its ability to retard the movement of radionuclides. Radionuclides that are transported through the geological media, to emerge at the bottom of the seawater volume, would then be subjected to the same processes of dilution, dispersion, diffusion, and sorption that affect radioactive waste disposed of at sea. This method of disposal therefore provides additional containment of radionuclides when compared with the disposal of wastes directly to the seabed.

Disposal in Ice Sheets

Since 1980 there has been no significant consideration of this option. Containers of heat-generating waste would be placed in stable ice sheets such as those found in Greenland and Antarctica. The containers would melt the surrounding ice and be drawn deep into the ice sheet, where the ice would refreeze above the wastes creating a thick barrier. Although disposal in ice sheets could be technically considered for all types of radioactive wastes, it has only been seriously investigated for HLW, where the heat generated by the wastes could be used to achieve self-burial within the ice by melting.

The option of disposal in ice sheets has not been implemented anywhere. It has been rejected by countries that have signed the 1959 Antarctic Treaty or have committed to providing a solution to their radioactive waste management within their national boundaries.

Deep Well Injection (Liquid)

This approach involves the injection of liquid radioactive waste directly into a layer of rock deep underground that has been chosen because of its suitable characteristics to trap the waste (*i.e.* minimise any further movement following injection).

In order to achieve this there are two geological prerequisites. There must be a layer of rock (injection layer) with sufficient porosity to accommodate the waste and with sufficient permeability to allow easy injection (*i.e.* act like a sponge). Above and below the injection layer there must be impermeable layers that act as a natural seal. Additional benefits could be provided from geological features that limit horizontal or vertical migration. For example, injection into layers of rock containing natural brine groundwater. This is because the high density of brine (salt water) would reduce the potential for upward movement.

Direct injection could in principle be used on any type of radioactive waste provided that it could be transformed into a solution or slurry (very fine particles in water). Slurries containing a cement grout that would set as a solid when underground could also be used to help minimise movement of radioactive waste. Direct injection has been implemented in Russia and the USA.

In 1957 extensive geological investigations started in Russia for suitable injection layers for radioactive waste. Three sites were found, all in sedimentary rocks. At Krasnoyarsk-26 and Tomsk-7 injection takes place into two porous sandstone beds capped by clays at depths up to 400 metres. Whereas at Dimitrovgrad injection has now stopped, but took place into sandstone and limestone formations at a depth of 1400 metres. In total, some tens of millions of cubic metres of LLW, ILW and HLW have been injected in Russia.

In the USA, direct injection of about 7500 cubic metres of LLW as cement slurries was undertaken during the 1970s at a depth of about 300 metres over a period of 10 years at the Oak Ridge National Laboratory, Tennessee. It was abandoned because of uncertainties over the migration of the grout in the surrounding fractured rocks (shales). In addition a scheme involving HLW injection into crystalline bedrock beneath the Savannah River Site in South Carolina was abandoned before it was implemented due to public concerns.

Tenorm

Radioactive material is produced or collected as a waste product from the oil and gas industry and generally referred to as ‘technologically enhanced naturally occurring radioactive material' (Tenorm). In oil and gas production, radium-226, radium-228 and lead-210 are deposited as scale in pipes and equipment in many parts of the world. Published data show radionuclide concentrations in scales up to 300,000 Bq/kg for Pb-210, 250,000 Bq/kg for Ra-226 and 100,000 Bq/kg for Ra-228. However, scrap steel from gas plants may be recycled if it has less than 500,000 Bq/kg (0.5 MBq/kg) radioactivity (the exemption level). This level however is 1000 times higher than the clearance level for recycled material (both steel and concrete) from the nuclear industry, where anything above 500 Bq/kg may not be cleared from regulatory control for recycling and must be disposed of, usually as intermediate-level waste.

The largest Tenorm waste stream is coal ash, with 280 million tonnes arising globally each year, and carrying uranium-238 and all its non-gaseous decay products, as well as thorium-232 and its progeny. This is usually just buried.

The double standard means that the same radionuclide, at the same concentration, can either be sent to deep disposal as waste (if scrap from the nuclear industry) or released for use in building materials (if contained in fly ash or recycled steel from oil/gas production).

References

- Nuclear-reactors: lumenlearning.com, Retrieved 2 July, 2019
- Nuclear-chain-reaction: energyeducation.ca, Retrieved 22 March, 2019
- Nuclear-fuel-cycle: world-nuclear.org, Retrieved 28 May, 2019
- Fuel-fabrication, conversion-enrichment-and-fabrication, nuclear-fuel-cycle: world-nuclear.org, Retrieved 14 January, 2019
- Fission-products-chemistry-in-sfrs, oxide-chemistry-in-sfrs, reactor-physics-and-nuclear-materials, radiation-science-technology: tudelft.nl, Retrieved 1 April, 2019
- Featurepwr-and-bwr-chemistry-optimization: neimagazine.com, Retrieved 19 June, 2019
- Storage-and-disposal-of-radioactive-wastes, nuclear-wastes, nuclear-fuel-cycle: world-nuclear.org, Retrieved 28 February, 2019

6

Nuclear Medicine

The branch of medicine which seeks to apply radioactive substances for diagnosing and treating diseases is called nuclear medicine. Some of its applications include nuclear medicine imaging and radiopharmaceuticals. The topics elaborated in this chapter will help in gaining a better perspective about these aspects and applications of nuclear medicine.

Nuclear medicine imaging uses small amounts of radioactive material to diagnose, evaluate or treat a variety of diseases. These include many types of cancers, heart disease, gastrointestinal, endocrine or neurological disorders and other abnormalities. Because nuclear medicine exams can pinpoint molecular activity, they have the potential to identify disease in its earliest stages. They can also show whether a patient is responding to treatment.

Diagnosis

Nuclear medicine imaging procedures are noninvasive. With the exception of intravenous injections, they are usually painless. These tests use radioactive materials called radiopharmaceuticals or radiotracers to help doctors diagnose and evaluate medical conditions.

Radiotracers are molecules linked to, or "labeled" with, a small amount of radioactive material that can be detected on the PET scan. Radiotracers accumulate in tumors or regions of inflammation. They can also bind to specific proteins in the body. The most commonly used radiotracer is F-18 fluorodeoxyglucose, or FDG, a molecule similar to glucose. Cancer cells are more metabolically active and may absorb glucose at a higher rate. This higher rate can be seen on PET scans. This allows the doctor to identify disease before it may be seen on other imaging tests. FDG is just one of many radiotracers in use or in development.

Depending on the type of exam, the radiotracer is injected, swallowed or inhaled as a gas. It eventually accumulates in the area of the body under examination. A special camera or imaging device detects radioactive emissions from the radiotracer. The camera or device produces pictures and provides molecular information.

Many centers superimpose nuclear medicine images with computed tomography (CT) or magnetic resonance imaging (MRI) to produce special views. This is known as image fusion or co-registration. These views allow the doctor to correlate and interpret information from two different exams on one image this leads to more precise information and accurate diagnoses. Single photon emission computed tomography/computed tomography (SPECT/CT) and positron emission tomography/computed tomography (PET/CT) units can perform both exams at the same time. PET/MRI is an emerging imaging technology. However, it is not universally available at this time.

Therapy

Nuclear medicine also offers therapeutic procedures, such as radioactive iodine (I-131) therapy that use small amounts of radioactive material to treat cancer and other medical conditions affecting the thyroid gland, as well as treatments for other cancers and medical conditions.

Non-Hodgkin's lymphoma patients who do not respond to chemotherapy may undergo radioimmunotherapy (RIT).

Radioimmunotherapy (RIT) is a personalized cancer treatment that combines radiation therapy with the targeting ability of immunotherapy, a treatment that mimics cellular activity in the body's immune system.

Common uses of the Procedure

Physicians use nuclear medicine imaging procedures to visualize the structure and function of an organ, tissue, bone or system within the body.

In adults, nuclear medicine is used to:

Heart

- Visualize heart blood flow and function (such as a myocardial perfusion scan),
- Detect coronary artery disease and the extent of coronary stenosis,
- Assess damage to the heart following a heart attack,
- Evaluate treatment options such as bypass heart surgery and angioplasty,
- Evaluate the results of revascularization (blood flow restoration) procedures,
- Detect heart transplant rejection,
- Evaluate heart function before and after chemotherapy (MUGA).

Lungs

- Scan lungs for respiratory and blood flow problems,
- Assess differential lung function for lung reduction or transplant surgery,
- Detect lung transplant rejection.

Bones

- Evaluate bones for fractures, infection and arthritis,
- Evaluate for metastatic bone disease,
- Evaluate painful prosthetic joints,
- Evaluate bone tumors,
- Identify sites for biopsy.

Brain

- Investigate abnormalities in the brain in patients with certain symptoms or disorders, such as seizures, memory loss and suspected abnormalities in blood flow,
- Detect the early onset of neurological disorders such as alzheimer's disease,
- Assist in surgical planning and identify the areas of the brain that may be causing seizures,
- Evaluate for abnormalities in a chemical in the brain involved in controlling movement in patients with suspected parkinson's disease or related movement disorders,
- Evaluation for suspected brain tumor recurrence, surgical or radiation planning or localization for biopsy.

Other Systems

- Identify inflammation or abnormal function of the gallbladder,
- Identify bleeding into the bowel,
- Assess post-operative complications of gallbladder surgery,
- Evaluate lymphedema,
- Evaluate fever of unknown origin,
- Locate the presence of infection,
- Measure thyroid function to detect an overactive or underactive thyroid,
- Help diagnose hyperthyroidism and blood cell disorders,
- Evaluate for hyperparathyroidism (overactive parathyroid gland),
- Evaluate stomach emptying,
- Evaluate spinal fluid flow and potential spinal fluid leaks.

In adults and children, nuclear medicine is also used to:

Cancer

- Stage cancer by determining the presence or spread of cancer in various parts of the body,
- Localize sentinel lymph nodes before surgery in patients with breast cancer or skin and soft tissue tumors,
- Plan treatment,
- Evaluate response to therapy,
- Detect the recurrence of cancer,
- Detect rare tumors of the pancreas and adrenal glands.

Renal

- Analyze native and transplant kidney blood flow and function,
- Detect urinary tract obstruction,
- Evaluate for hypertension (high blood pressure) related to the kidney arteries,
- Evaluate kidneys for infection versus scar,
- Detect and follow-up urinary reflux.

In children, nuclear medicine is also used to:

- Investigate abnormalities in the esophagus, such as esophageal reflux or motility disorders,
- Evaluate the openness of tear ducts,
- Evaluate the openness of ventricular shunts in the brain,
- Assess congenital heart disease for shunts and pulmonary blood flow.

Nuclear medicine therapies include:

- Radioactive iodine (I-131) therapy used to treat some causes of hyperthyroidism (overactive thyroid gland, for example, Graves' disease) and thyroid cancer,
- Radioactive antibodies used to treat certain forms of lymphoma (cancer of the lymphatic system),
- Radioactive phosphorus (P-32) used to treat certain blood disorders,
- Radioactive materials used to treat painful tumor metastases to the bones,

- I-131 MIBG (radioactive iodine labeled with metaiodobenzylguanidine) used to treat adrenal gland tumors in adults and adrenal gland/nerve tissue tumors in children.

What does the Equipment Look Like?

The special camera and imaging techniques used in nuclear medicine include the gamma camera and single-photon emission-computed tomography (SPECT).

The gamma camera, also called a scintillation camera, detects radioactive energy that is emitted from the patient's body and converts it into an image. The gamma camera itself does not emit any radiation. The gamma camera is composed of radiation detectors, called gamma camera heads, which are encased in metal and plastic and most often shaped like a box, attached to a round circular donut shaped gantry. The patient lies on the examination table which slides in between two parallel gamma camera heads that are positioned above the patient. Sometimes, the gamma camera heads are oriented at a 90 degree angle and placed over the patient's body.

SPECT involves the rotation of the gamma camera heads around the patient's body to produce more detailed, three-dimensional images.

A PET scanner is a large machine with a round, donut-shaped hole in the middle. It looks similar to a CT or MRI unit. Multiple rings of detectors inside the machine record the energy emissions from the radiotracer in the body.

A computer helps create the images from the data obtained by the gamma camera.

A probe is a small hand-held device resembling a microphone that can detect and measure the amount of the radiotracer in a small area of the body.

There is no specialized equipment used during radioactive iodine therapy, but the technologist or other personnel administering the treatment may cover the clothing and use lead containers to shield the radioactive material you will be receiving.

How does the Procedure Work?

Ordinary x-ray exams create an image by passing x-rays through the body. Nuclear medicine exams use a radioactive material called a radiopharmaceutical or radiotracer. This material is injected into the bloodstream, swallowed or inhaled as a gas. The material accumulates in the area of the body under examination, where it gives off a small amount of energy in the form of gamma rays. Special cameras detect this energy and, with the help of a computer, create pictures that offer details on the structure and function of organs and tissues.

Unlike other imaging techniques, nuclear medicine exams focus on processes within the body, such as rates of metabolism or levels of various other chemical activity. Areas of greater intensity, called "hot spots," indicate where large amounts of the radiotracer have accumulated and where there is a high level of chemical or metabolic activity. Less intense areas, or "cold spots," indicate a smaller concentration of radiotracer and less activity.

In radioactive iodine (I-131) therapy for thyroid disease, radioactive iodine (I-131) is swallowed, absorbed into the bloodstream in the gastrointestinal (GI) tract and absorbed from the blood by the thyroid gland where it destroys cells within that organ.

Radioimmunotherapy (RIT) is a combination of radiation therapy and immunotherapy. In immunotherapy, a laboratory-produced molecule called a monoclonal antibody is engineered to recognize and bind to the surface of cancer cells. Monoclonal antibodies mimic the antibodies naturally produced by the body's immune system that attack invading foreign substances, such as bacteria and viruses.

In RIT, a monoclonal antibody is paired with a radioactive material. When injected into the patient's bloodstream, the antibody travels to and binds to the cancer cells, allowing a high dose of radiation to be delivered directly to the tumor.

In I-131MIBG therapy for neuroblastoma, the radiotracer is administered by injection into the blood stream. The radiotracer binds to the cancer cells allowing a high dose of radiation to be delivered to the tumor.

Benefits

- Nuclear medicine examinations provide unique information—including details on the function and anatomy of body structures—that is often unattainable using other imaging procedures.
- Nuclear medicine scans provide the most useful diagnostic or treatment information for many diseases.
- A nuclear medicine scan is less expensive and may yield more precise information than exploratory surgery.
- Nuclear medicine offers the potential to identify disease in its earliest stage, often before symptoms occur or abnormalities can be detected with other diagnostic tests.
- By detecting whether lesions are likely benign or malignant, PET scans may eliminate the need for surgical biopsy or identify the best biopsy location.
- PET scans may provide additional information that is used for radiation therapy planning.

Risks

- Because only a small dose of radiotracer is used, nuclear medicine exams have a relatively low radiation exposure. This is acceptable for diagnostic exams. Thus, the radiation risk is very low when compared with the potential benefits.

- Nuclear medicine diagnostic procedures have been used for more than five decades, and there are no known long-term adverse effects from such low-dose exposure.
- Treatment risks are always weighed against the potential benefits for nuclear medicine therapeutic procedures. The doctor will inform you of all significant risks prior to the treatment and give you an opportunity to ask questions.
- Allergic reactions to radiotracers are extremely rare and usually mild. Always tell the nuclear medicine personnel of any allergies you may have or other problems that may have occurred during a previous nuclear medicine exam.
- Injection of the radiotracer may cause slight pain and redness. This should rapidly resolve.
- Women should always tell their doctor and radiology technologist if there is any possibility that they are pregnant or they are breastfeeding.

What are the Limitations of General Nuclear Medicine?

Nuclear medicine procedures can be time consuming. It can take several hours to days for the radiotracer to accumulate in the area of interest, and imaging may take up to several hours to perform. In some cases, newer equipment can substantially shorten the procedure time.

The image resolution of nuclear medicine images may not be as high as that of CT or MRI. However, nuclear medicine scans are more sensitive for a variety of indications, and the functional information they yield is often unobtainable by other imaging techniques.

Nuclear Medicine Imaging

Nuclear medicine imaging is a method of producing images by detecting radiation from different parts of the body after a radioactive tracer is given to the patient. The images are digitally generated on a computer and transferred to a nuclear medicine physician, who interprets the images to make a diagnosis.

Radioactive tracers used in nuclear medicine are, in most cases, injected into a vein. For some studies, they may be given by mouth. These tracers are not dyes or medicines, and they have no side effects. The amount of radiation a patient receives in a typical nuclear medicine scan tends to be very low.

Nuclear imaging produces images by detecting radiation from different parts of the body after a radioactive tracer material is administered. The images are recorded on computer and on film. Different types of nuclear imaging tests have different preparation instructions.

Types of Nuclear Imaging

Types of nuclear imaging include:

- Bone density scan: An enhanced form of X-ray technology that is used to measure bone loss or density.
- Cardiac PET perfusion: An evaluation of the blood flow (perfusion) to the walls of your heart using a high resolution PET scanner. Usually performed using a cardiac stress test.
- Cardiac PET sarcoid: Similar to Cardiac PET Viability except with different eating instructions prior to the exam. An evaluation of the functional status of the heart (viability) and whether the heart has suffered permanent damage from sarcoidosis.
- Cardiac PET viability: An evaluation of the functional status of the heart (viability) and whether the heart has suffered permanent damage.
- Cardiac SPECT perfusion: An evaluation of the blood flow (perfusion) to the walls of your heart. Usually performed using a cardiac stress test.
- PET/CT scanning: Combined PET/CT scans provide images that pinpoint the location of abnormal metabolic activity within the body. The combined scans have been shown to provide more accurate diagnoses than the two scans performed separately.

Radioisotopes in Medicine

Radioactive isotopes, or radioisotopes, are species of chemical elements that are produced through the natural decay of atoms. Exposure to radiation generally is considered harmful to the human body, but radioisotopes are highly valuable in medicine, particularly in the diagnosis and treatment of disease.

Nuclear medicine uses radioactive isotopes in a variety of ways. One of the more common uses is as a tracer in which a radioisotope, such as technetium-99m, is taken orally or is injected or is inhaled into the body. The radioisotope then circulates through the body or is taken up only by certain tissues. Its distribution can be tracked according to the radiation it gives off. The emitted radiation can be captured by various imaging techniques, such as single photon emission computed tomography (SPECT) or positron emission tomography (PET), depending on the radioisotope used. Through such imaging, physicians are able to examine blood flow to specific organs and assess organ function or bone growth. Radioisotopes typically have short half-lives and typically decay before their emitted radioactivity can cause damage to the patient's body.

Therapeutic applications of radioisotopes typically are intended to destroy the targeted cells. This approach forms the basis of radiotherapy, which is commonly used to treat cancer and other conditions involving abnormal tissue growth, such as hyperthyroidism. In radiation therapy for cancer, the patient's tumor is bombarded with ionizing radiation, typically in the form of beams of subatomic particles, such as protons, neutrons, or alpha or beta particles, which directly disrupt the atomic or molecular structure of the targeted tissue. Ionizing radiation introduces breaks in the double-stranded DNA molecule, causing the cancer cells to die and thereby preventing their replication. While radiotherapy is associated with unpleasant side effects, it generally is effective in slowing cancer progression or, in some cases, even prompting the regression of malignant disease.

The use of radioisotopes in the fields of nuclear medicine and radiotherapy has advanced significantly since the discovery of artificial radioisotopes in the first decades of the 1900s. Artificial radioisotopes are produced from stable elements that are bombarded with neutrons. Following that discovery, researchers began to investigate potential medical applications of artificial radioisotopes, work that laid the foundation for nuclear medicine. Today diagnostic and therapeutic procedures using radioactive isotopes are routine.

Uses of Radioisotopes

Radioactive isotopes have the same chemical properties as stable isotopes of the same element, but they emit radiation, which can be detected. If we replace one (or more) atom(s) with radioisotope(s) in a compound, we can track them by monitoring their radioactive emissions. This type of compound is called a radioactive tracer (or radioactive label). Radioisotopes are used to follow the paths of biochemical reactions or to determine how a substance is distributed within an organism. Radioactive tracers are also used in many medical applications, including both diagnosis and treatment. They are used to measure engine wear, analyze the geological formation around oil wells, and much more.

Radioisotopes have revolutionized medical practice, where they are used extensively. Over 10 million nuclear medicine procedures and more than 100 million nuclear medicine tests are performed annually in the United States. Four typical examples of radioactive tracers used in medicine are technetium-99 ($^{99}_{43}Tc$), thallium-201 ($^{201}_{81}Tl$), iodine-131 ($^{131}_{53}I$), and sodium-24 ($^{24}_{11}Na$). Damaged tissues in the heart, liver, and lungs absorb certain compounds of technetium-99 preferentially. After it is injected, the location of the technetium compound, and hence the damaged tissue, can be determined by detecting the γ rays emitted by the Tc-99 isotope. Thallium-201 becomes concentrated in healthy heart tissue, so the two isotopes, Tc-99 and Tl-201, are used together to study heart tissue. Iodine-131 concentrates in the thyroid gland, the liver, and some parts of the brain. It can therefore be used to monitor goiter and treat thyroid conditions, such as Grave's disease, as well as liver and brain tumors. Salt solutions containing compounds of sodium-24 are injected into the bloodstream to help locate obstructions to the flow of blood.

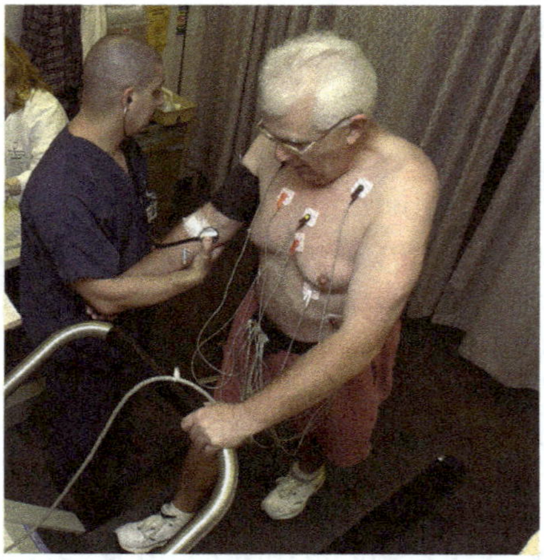

Administering thallium-201 to a patient and subsequently performing a stress test offer medical professionals an opportunity to visually analyze heart function and blood flow.

Radioisotopes used in medicine typically have short half-lives—for example, the ubiquitous Tc-99m has a half-life of 6.01 hours. This makes Tc-99m essentially impossible to store and prohibitively expensive to transport, so it is made on-site instead. Hospitals and other medical facilities use Mo-99 (which is primarily extracted from U-235 fission products) to generate Tc-99. Mo-99 undergoes β decay with a half-life of 66 hours, and the Tc-99 is then chemically extracted. The parent nuclide Mo-99 is part of a molybdate ion, MoO_4^{2-}; when it decays, it forms the pertechnetate ion, TcO_4^-. These two water-soluble ions are separated by column chromatography, with the higher charge molybdate ion adsorbing onto the alumina in the column, and the lower charge pertechnetate ion passing through the column in the solution. A few micrograms of Mo-99 can produce enough Tc-99 to perform as many as 10,000 tests.

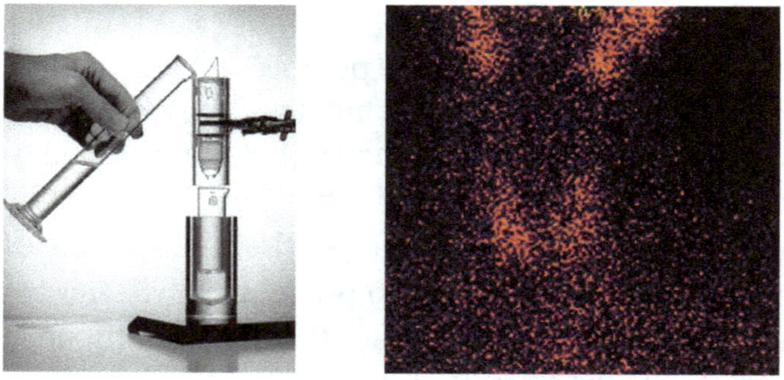

(a) (b)

Figure: (a) The first Tc-99m generator (circa 1958) is used to separate Tc-99 from Mo-99. The MoO_4^{2-} is retained by the matrix in the column, whereas the TcO_4^-. passes through and is collected. (b) Tc-99 was used in this scan of the neck of a patient with Grave's disease. The scan shows the location of high concentrations of Tc-99.

Radioisotopes can also be used, typically in higher doses than as a tracer, as treatment. Radiation therapy is the use of high-energy radiation to damage the DNA of cancer cells, which kills them or keeps them from dividing. A cancer patient may receive external beam radiation therapy delivered by a machine outside the body, or internal radiation therapy (brachytherapy) from a radioactive substance that has been introduced into the body. Note that chemotherapy is similar to internal radiation therapy in that the cancer treatment is injected into the body, but differs in that chemotherapy uses chemical rather than radioactive substances to kill the cancer cells.

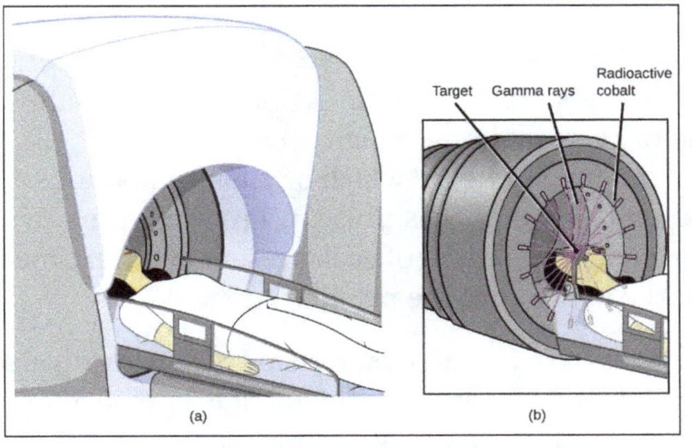

The diagram above in (a) Shows a cobalt-60 machine used in the treatment of cancer. The diagram in (b) Shows how the gantry of the Co-60 machine swings through an arc, focusing radiation on the targeted region (tumor) and minimizing the amount of radiation that passes through nearby regions.

Cobalt-60 is a synthetic radioisotope produced by the neutron activation of Co-59, which then undergoes β decay to form Ni-60, along with the emission of γ radiation. The overall process is:

$$^{59}_{27}\mathrm{Co} + ^{1}_{0}\mathrm{n} \rightarrow ^{60}_{27}\mathrm{Co} \rightarrow ^{60}_{28}\mathrm{Ni} + ^{0}_{-1}\beta + 2^{0}_{0}\gamma$$

The overall decay scheme for this is shown graphically.

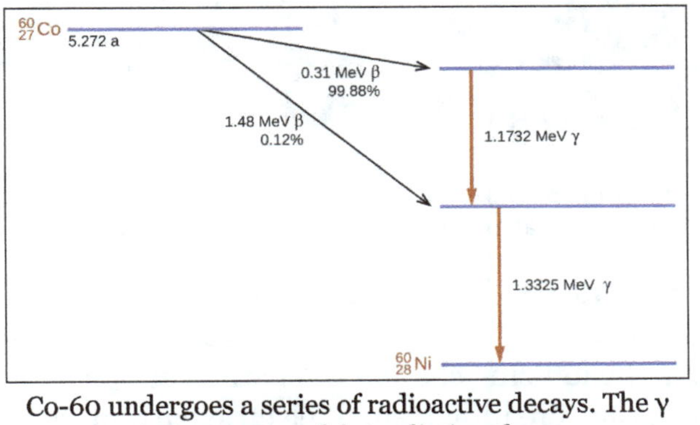

Co-60 undergoes a series of radioactive decays. The γ emissions are used for radiation therapy.

Radioisotopes are used in diverse ways to study the mechanisms of chemical reactions in plants and animals. These include labeling fertilizers in studies of nutrient uptake by plants and crop growth, investigations of digestive and milk-producing processes in cows, and studies on the growth and metabolism of animals and plants.

For example, the radioisotope C-14 was used to elucidate the details of how photosynthesis occurs. The overall reaction is:

$$6CO_2(g)+6H_2O(l)\rightarrow C_6H_{12}O_6(s)+6O(g),$$

But the process is much more complex, proceeding through a series of steps in which various organic compounds are produced. In studies of the pathway of this reaction, plants were exposed to CO_2 containing a high concentration of ${}^{14}_{4}C$. At regular intervals, the plants were analyzed to determine which organic compounds contained carbon-14 and how much of each compound was present. From the time sequence in which the compounds appeared and the amount of each present at given time intervals, scientists learned more about the pathway of the reaction.

Commercial applications of radioactive materials are equally diverse. They include determining the thickness of films and thin metal sheets by exploiting the penetration power of various types of radiation. Flaws in metals used for structural purposes can be detected using high-energy gamma rays from cobalt-60 in a fashion similar to the way X-rays are used to examine the human body. In one form of pest control, flies are controlled by sterilizing male flies with γ radiation so that females breeding with them do not produce offspring. Many foods are preserved by radiation that kills microorganisms that cause the foods to spoil.

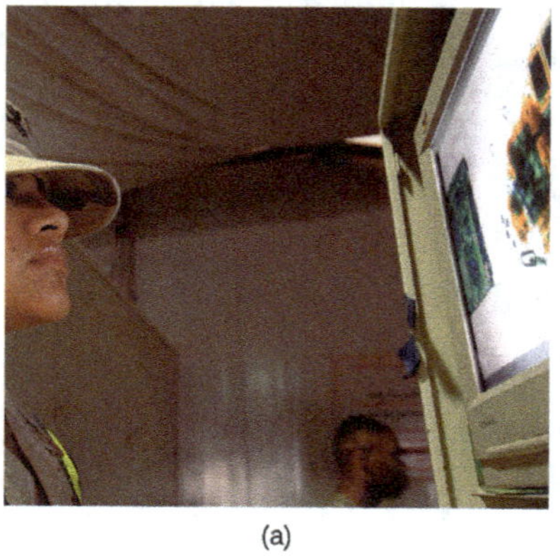

(a)

(b)

Common commercial uses of radiation include (a) X-ray examination of luggage at an airport and (b) Preservation of food.

Americium-241, an α emitter with a half-life of 458 years, is used in tiny amounts in ionization-type smoke detectors. The α emissions from Am-241 ionize the air between two electrode plates in the ionizing chamber. A battery supplies a potential that causes movement of the ions, thus creating a small electric current. When smoke enters the chamber, the movement of the ions is impeded, reducing the conductivity of the air. This causes a marked drop in the current, triggering an alarm.

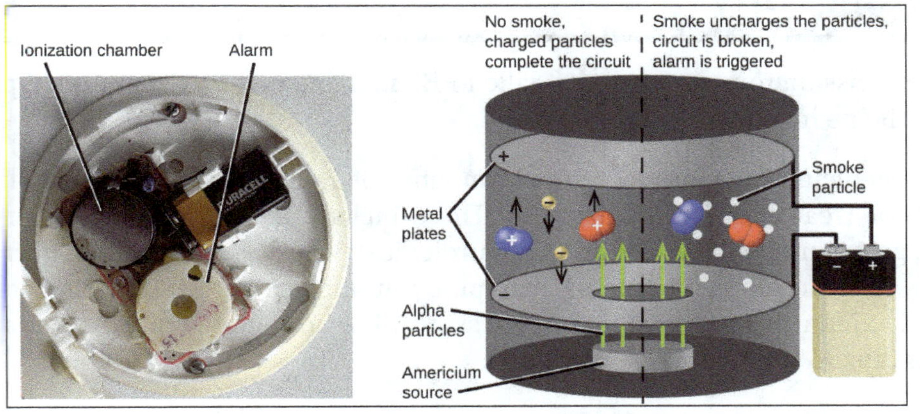

Inside a smoke detector, Am-241 emits α particles that ionize the air, creating a small electric current. During a fire, smoke particles impede the flow of ions, reducing the current and triggering an alarm.

Radiopharmaceuticals

Radiopharmaceuticals are radioisotopes bound to biological molecules able to target specific organs, tissues or cells within the human body. These radioactive drugs can be used for the diagnosis and, increasingly, for the therapy of diseases.

The number of radiopharmaceuticals in clinical use is rapidly growing, thus allowing the medical community better access to detail information on the characteristics of the different types of tumours.

A radiopharmaceutical can be seen as an entity made up of a radionuclide and a vehicle molecule with high affinity – or binding power – for a tissue or a specific function of a human organ. It may also comprise only the radioisotope itself if it shows suitable biological properties.

Radiopharmaceuticals are used to produce images of organs or tissues of interest, a process that is called scintigraphy. A type of medical device known as gamma camera is able to detect the gamma rays emitted by the radioisotope. It generates in a non-invasive manner images that reflect the function of the organ or tissue under investigation.

The most widely used radioisotope in diagnostic nuclear medicine is technetium-99m. It can be attached to several specific molecules, allowing the diagnosis of many diseases, including certain types of cancers. For instance, technetium-99m-MDP (methylene diphosphonate) is widely used to detect bone metastasis associated with cancer.

Positron Emission Tomography - Computed Tomography

Positron emission tomography, also called PET imaging or a PET scan, is a type of nuclear medicine imaging.

Nuclear medicine imaging uses small amounts of radioactive material to diagnose, evaluate or treat a variety of diseases. These include many types of cancers, heart disease, gastrointestinal, endocrine or neurological disorders and other abnormalities. Because nuclear medicine exams can pinpoint molecular activity, they have the potential to identify disease in its earliest stages. They can also show whether a patient is responding to treatment.

Nuclear medicine imaging procedures are noninvasive. With the exception of intravenous injections, they are usually painless. These tests use radioactive materials called radiopharmaceuticals or radiotracers to help doctors diagnose and evaluate medical conditions.

Radiotracers are molecules linked to, or "labeled" with, a small amount of radioactive material that can be detected on the PET scan. Radiotracers accumulate in tumors or regions of inflammation. They can also bind to specific proteins in the body. The most commonly used radiotracer is F-18 fluorodeoxyglucose, or FDG, a molecule similar to glucose. Cancer cells are more metabolically active and may absorb glucose at a higher rate. This higher rate can be seen on PET scans. This allows your doctor to identify disease before it may be seen on other imaging tests. FDG is just one of many radiotracers in use or in development.

Depending on the type of exam, the radiotracer is injected, swallowed or inhaled as a gas. It eventually accumulates in the area of the body under examination. A special camera or imaging device detects radioactive emissions from the radiotracer. The camera or device produces pictures and provides molecular information.

Many centers superimpose nuclear medicine images with computed tomography (CT) or magnetic resonance imaging (MRI) to produce special views. This is known as image fusion or co-registration. These views allow the doctor to correlate and interpret information from two different exams on one image. This leads to more precise information and accurate diagnoses. Single photon emission computed tomography/computed tomography (SPECT/CT) and positron emission tomography/computed tomography (PET/CT) units can perform both exams at the same time. PET/MRI is an emerging imaging technology. However, it is not universally available at this time.

A PET scan measures important body functions, such as metabolism. It helps doctors evaluate how well organs and tissues are functioning. CT imaging uses special x-ray equipment, and in some cases a contrast material, to produce multiple images of the inside of the body. A radiologist views and interprets these images on a computer monitor. CT imaging provides excellent anatomic information.

Combined PET/CT scanners perform almost all PET scans today. These combined scans help pinpoint abnormal metabolic activity and may provide more accurate diagnoses than the two scans performed separately.

Common uses of the Procedure

Doctors perform PET and PET/CT scans to:

- Detect cancer and/or make a diagnosis.
- Determine whether a cancer has spread in the body.
- Assess the effectiveness of treatment.
- Determine if a cancer has returned after treatment.
- Evaluate prognosis.
- Assess tissue metabolism and viability.
- Determine the effects of a heart attack myocardial infarction on areas of the heart.
- Identify areas of the heart muscle that would benefit from angioplasty or coronary artery bypass surgery (in combination with a myocardial perfusion scan).
- Evaluate brain abnormalities, such as tumors, memory disorders, seizures and other central nervous system disorders.
- Map normal human brain and heart function.

What does the Equipment Look Like?

A PET scanner is a large machine with a round, donut-shaped hole in the middle. It looks similar to a CT or MRI unit. Multiple rings of detectors inside the machine record the energy emissions from the radiotracer in your body.

The CT scanner is typically a large, donut-shaped machine with a short tunnel in the center. You will lie on a narrow examination table that slides in and out of this short tunnel. Rotating around you, the X-ray tube and electronic x-ray detectors are located opposite each other in a ring, called a gantry. The computer workstation that processes the imaging information is located in a separate control room. This is where the technologist operates the scanner and monitors your exam in direct visual contact. The technologist will be able to hear and talk to you using a speaker and microphone. Combined PET/CT scanners look similar to both the PET and CT scanners. A computer helps create the images from the data obtained by the gamma camera.

How does the Procedure Work?

Ordinary X-ray exams create an image by passing x-rays through the body. Nuclear medicine exams use a radioactive material called a radiopharmaceutical or radiotracer. This material is injected into the bloodstream, swallowed or inhaled as a gas. The material accumulates in the area of your body under examination, where it gives off a small amount of energy in the form of gamma rays. Special cameras detect this energy and, with the help of a computer, create pictures that offer details on the structure and function of organs and tissues. PET scans only use radiotracer injections.

Unlike other imaging techniques, nuclear medicine exams focus on processes within the body, such as rates of metabolism or levels of various other chemical activity. Areas of greater intensity, called "hot spots," indicate where large amounts of the radiotracer have accumulated and where there is a high level of chemical or metabolic activity. Less intense areas, or "cold spots," indicate a smaller concentration of radiotracer and less activity.

Benefits

- Nuclear medicine examinations provide unique information—including details on the function and anatomy of body structures—that is often unattainable using other imaging procedures.
- Nuclear medicine scans provide the most useful diagnostic or treatment information for many diseases.
- A nuclear medicine scan is less expensive and may yield more precise information than exploratory surgery.
- By identifying changes in the body at the cellular level, PET imaging may detect the early onset of disease before it is evident on other imaging tests such as CT or MRI.

The benefits of a combined PET/CT scan include:

- Greater detail with a higher level of accuracy; because both scans are performed at the same time without the patient having to change positions, there is less room for error.
- Greater convenience for the patient who undergoes ct and pet at one time rather than two different times.

Risks

- Because only a small dose of radiotracer is used, nuclear medicine exams have a relatively low radiation exposure. This is acceptable for diagnostic exams. Thus, the radiation risk is very low when compared with the potential benefits.

- Nuclear medicine diagnostic procedures have been used for more than five decades, and there are no known long-term adverse effects from such low-dose exposure.
- Treatment risks are always weighed against the potential benefits for nuclear medicine therapeutic procedures. Your doctor will inform you of all significant risks prior to the treatment and give you an opportunity to ask questions.
- Allergic reactions to radiotracers are extremely rare and usually mild. Always tell the nuclear medicine personnel of any allergies you may have or other problems that may have occurred during a previous nuclear medicine exam.
- Injection of the radiotracer may cause slight pain and redness. This should rapidly resolve.
- Women should always tell their doctor and radiology technologist if there is any possibility that they are pregnant or they are breastfeeding.
- For risks of CT exams.

Limitations of Positron Emission Tomography - computed Tomography

Nuclear medicine procedures can be time consuming. It can take several hours to days for the radiotracer to accumulate in the area of interest, and imaging may take up to several hours to perform. In some cases, newer equipment can substantially shorten the procedure time.

The image resolution of nuclear medicine images may not be as high as that of CT or MRI. However, nuclear medicine scans are more sensitive for a variety of indications, and the functional information they yield is often unobtainable by other imaging techniques. Altered blood sugar or blood insulin levels may adversely affect the test results of diabetic patients or patients who have eaten a few hours prior to the exam.

The radiotracer decays quickly and is effective for only a short time. Therefore, it is important for you to be on time for your appointment and to receive the radioactive material at the scheduled time. Late arrival for an appointment may require you to reschedule the procedure. A very obese person may not fit into the opening of a conventional PET/CT unit.

PET Radiotracers

A PET radiotracer (also known as PET tracer) is a positron-emitting radiopharmaceutical used in positron emission tomography. Each tracer consists of a positron-emitting isotope (radioactive tag) bound to an organic ligand (targeting agent). The ligand component of each tracer interacts with a protein target, resulting in a characteristic distribution of the tracer throughout the tissues.

For example, the preeminent PET radiotracer fluorodeoxyglucose (FDG) is comprised of ^{18}F isotope bound to 2-deoxy-2-glucose, an analog to glucose. The 2-deoxy-2-glucose ligand is a substrate for the hexokinase/glucokinase enzymes involved in early carbohydrate metabolism; thus, FDG is chemically linked to cellular metabolic activity. It serves as a particularly good tracer agent because it tends to stay "trapped" within metabolically active cells due to the absence of the hydroxide group ("deoxy-").

There is an increasing list of chemical compounds which are being used for PET imaging. A list of commonly used compounds includes:

- Acetate (^{11}C)
- Choline (^{11}C)
- Fludeoxyglucose (^{18}F)
- Sodium fluoride (^{18}F)
- Methionine (^{11}C)
- Prostate-specific membrane antigen (PSMA) (^{68}Ga)
- DOTATOC, DOTANOC, DOTATATE (^{68}Ga)
- Florbetaben, Florbetapir (^{18}F)
- Rubidium (^{82}Rb) chloride.

Nuclear Scintigraphy

Nuclear scintigraphy (sometimes known as "bone scanning") is an invaluable diagnostic tool in some lameness cases. It is particularly useful in cases where nerve blocks have not revealed the source of the lameness, when other imaging techniques have not shown any abnormalities, in cases of multi-limb lameness and in difficult to image areas such as the pelvis. It is mainly used for imaging bones, although vascular and soft tissue phases are also possible.

The procedure involves injecting the horse with a radioactive "dye" (technetium 99), which is taken up by areas of increased bone activity. Shortly after this a special camera is used to detect the radioactivity produced and provide detailed images of the skeleton. It is particularly useful to identify stress fractures, abnormal stress modelling and to monitor fracture healing. It is also very useful in assessing some cases of back and neck pain.

Although it is well known to be a safe procedure, after injection and imaging the horse is kept in controlled conditions for twenty four hours to comply with radiation safety regulation requirements. Horses are generally allowed to leave the hospital the day after imaging, but sometimes the technique identifies problems which require further investigation. This must be done the next day to avoid unnecessary handling when the radioactivity levels are relatively high.

Radiotherapy

Radiotherapy uses high-energy rays, such as x-rays, to treat cancer. It destroys cancer cells in the area where it is given. Some normal cells in the area can also be damaged by radiotherapy. This can cause side effects. These normal cells are usually able to repair themselves, but cancer cells cannot. As the normal cells recover, the side effects usually get better.

Radiotherapy is always carefully planned by a team of experts. Radiotherapy treatment is continuing to become even more accurate. This allows your team to treat the cancer more effectively, while doing as little harm as possible to normal cells.

Many people have radiotherapy as part of their cancer treatment. Radiotherapy is given for different reasons.

Radiotherapy to Cure the Cancer

This is when radiotherapy is given to try to destroy a tumour and cure the cancer. It is sometimes called radical treatment. You may have radiotherapy on its own or sometimes with chemotherapy (chemoradiation).

Radiotherapy can be given after surgery for some types of cancer, to lower the risk of the cancer coming back. This is called adjuvant radiotherapy.

Sometimes radiotherapy is given before surgery to shrink a cancer and make it easier to remove. You may have it along with chemotherapy. This is sometimes called neo-adjuvant radiotherapy.

Radiotherapy to Treat Symptoms (Palliative Radiotherapy)

This is when radiotherapy is given to shrink the cancer and help with symptoms such as pain or coughing. Doctors may suggest this treatment to help control the cancer for longer when curing the cancer is not possible.

Ways of Giving Radiotherapy

There are two ways of giving radiotherapy:

- External beam radiotherapy is given from outside the body (externally) by a radiotherapy machine.
- Internal radiotherapy is when a radioactive material is placed inside the body. It is sometimes called brachytherapy or radioisotope therapy.

How you are given radiotherapy will depend on the type of cancer you have and where it is in the body. Some cancers are treated with both external and internal radiotherapy. Radiotherapy treatment is planned carefully for each person. This means that even if you know someone with the same type of cancer as you, their radiotherapy treatment may be different.

Chemoradiation

Sometimes chemotherapy is given at the same time as radiotherapy. This is called chemoradiation or chemoradiotherapy. Chemotherapy uses anti-cancer (cytotoxic) drugs to destroy cancer cells. The chemotherapy drugs can make cancer cells more sensitive to radiotherapy. This can help the radiotherapy to work better. This is only helpful for certain types of cancer, so it is not suitable for everyone. Giving chemotherapy and radiotherapy together can make the side effects of treatment worse.

References

- Gennuclear: radiologyinfo.org, Retrieved 3 February, 2019
- Nuclear-medicine-imaging: clevelandclinic.org, Retrieved 14 July, 2019
- Types-tests, nuclear-medicine-molecular-imaging: stanfordhealthcare.org, Retrieved 11 April, 2019
- How-radioactive-isotopes-are-used-in-medicine: britannica.com, Retrieved 10 January, 2019
- Uses-of-radioisotopes: opentextbc.ca, Retrieved 20 May, 2019
- Diagnostic-radiopharmaceuticals: iaea.org, Retrieved 29 August, 2019
- Pet-radiotracers: radiopaedia.org, Retrieved 16 January, 2019
- Nuclear-scintigraphy: randwickequine.com.au, Retrieved 26 March, 2019
- What-is-radiotherapy, radiotherapy: macmillan.org.uk, Retrieved 6 June, 2019

Permissions

All chapters in this book are published with permission under the Creative Commons Attribution Share Alike License or equivalent. Every chapter published in this book has been scrutinized by our experts. Their significance has been extensively debated. The topics covered herein carry significant information for a comprehensive understanding. They may even be implemented as practical applications or may be referred to as a beginning point for further studies.

We would like to thank the editorial team for lending their expertise to make the book truly unique. They have played a crucial role in the development of this book. Without their invaluable contributions this book wouldn't have been possible. They have made vital efforts to compile up to date information on the varied aspects of this subject to make this book a valuable addition to the collection of many professionals and students.

This book was conceptualized with the vision of imparting up-to-date and integrated information in this field. To ensure the same, a matchless editorial board was set up. Every individual on the board went through rigorous rounds of assessment to prove their worth. After which they invested a large part of their time researching and compiling the most relevant data for our readers.

The editorial board has been involved in producing this book since its inception. They have spent rigorous hours researching and exploring the diverse topics which have resulted in the successful publishing of this book. They have passed on their knowledge of decades through this book. To expedite this challenging task, the publisher supported the team at every step. A small team of assistant editors was also appointed to further simplify the editing procedure and attain best results for the readers.

Apart from the editorial board, the designing team has also invested a significant amount of their time in understanding the subject and creating the most relevant covers. They scrutinized every image to scout for the most suitable representation of the subject and create an appropriate cover for the book.

The publishing team has been an ardent support to the editorial, designing and production team. Their endless efforts to recruit the best for this project, has resulted in the accomplishment of this book. They are a veteran in the field of academics and their pool of knowledge is as vast as their experience in printing. Their expertise and guidance has proved useful at every step. Their uncompromising quality standards have made this book an exceptional effort. Their encouragement from time to time has been an inspiration for everyone.

The publisher and the editorial board hope that this book will prove to be a valuable piece of knowledge for students, practitioners and scholars across the globe.

Index

www.ingramcontent.com/pod-product-compliance
Lightning Source LLC
LaVergne TN
LVHW080608210926
848485LV00003B/123